The Six Sigma Yellowbelt Handbook

Paul Allen

Dedication

This book is dedicated to Paul Morris. A great friend and mentor that encouraged me to step into the world of consultancy and challenged me to deliver my best.
His sage advice over 5 years is still the wisdom that I draw on constantly with my clients and running my business. When I left his business to start on my own. He said 'Brilliant! How can I help!

To Paul Morris a man taken from this world far too early.

978-1-4710-2014-8
Imprint: Lulu.com

Contents

Contents

Introduction

25 years ago, I was selected to go on a World-Class Problem-Solving course. At the time it was called Six Sigma Blackbelt training. As part of the training I was given a technical problem to solve on a manufacturing process that I'd never seen before and one that I had no previous knowledge or experience of. Using simple principles and tools that I learned on the course and a team of Doers, operators, technicians etc, who regularly interacted with the process, 4 months later the technical problem had been fixed! WOW! In 20 previous years of my engineering career I had never had this capability. I soon realised in the next few years that this capability is simple to understand and deploy and although I can do some highly complex analysis it is the simple tools that solve these technical problems 95% of the time and that anyone can learn to do this and become a powerful money making problem solver.

I want you to have those same skills and see how simple it is to solve complex process problems. This problem-solving capability is described in this book and you are now on path to becoming world class... Welcome to your Six Sigma Yellowbelt Handbook...

Just to set the context of your role as a Yellowbelt. The role is about practical problem solving, being part of a team looking at an individual manufacturing process, potentially one of your local machines that is misbehaving. Perhaps, producing too many rejects or producing unreliable output quantities.

This book is designed to give you the skills to fix these types of problems in your work area. It is not designed to teach you the corporate arrangements and structure of using Six Sigma more broadly. That is fluff that will be of little value to practical local problem solving. I want you to have the skill to remove frustrating performance from your processes and make more money!

This book is also not about the statistical or mathematical process of creating any of the tools described in the text. But I guess a simple way to explain the intent is that when you attend a training course the lecturer could concentrate on teaching WHAT, WHY or HOW. In preparing to write this book I've read through some great stuff that teaches the HOW. How each graph and diagram is created.

But the thing that I want to concentrate on is the WHAT and the WHY you need to use these tools. That is what will really get you motivated to change your current approach. WHY am I going to do it a different way? WHY am I going to use more diagrams and statistical tools, etc? If you don't know WHY...then HOW really doesn't matter, as you're unlikely to want to be using the HOW anyway.

So we'll be setting the tools in context, inside a process. So that not only do you know how to use the individual tool but you know when to use the tools in real situations where you are trying to understand and fix a process problem.

Here are the quality tools we are recommending a Six Sigma Yellowbelt becomes proficient in using

Run chart
Measurement System Analysis
Histogram
Multi-vari Chart
Pareto Chart
Flow Diagram
Cause and Effect Diagram

Cpk Diagram

Control Charts and Statistical Process Control (SPC)

Run Chart

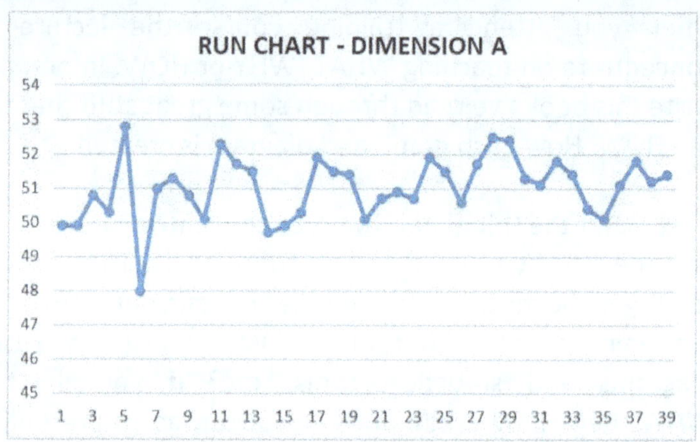

Measurement System Analysis

MSA ANOVA Method Results

Source	Variance	Standard Deviation	% Contribution
Total Measurement (Gage)	5.577E-07	0.00074679	49.33%
Repeatability	5.5255E-07	0.000743338	48.88%
Reproducibility	5.1446E-09	7.17256E-05	0.46%
Operator	5.1446E-09	7.17256E-05	0.46%
Oper * Part Interaction			
Product (Part-to-Part)	5.7282E-07	0.000756849	50.67%
Total	1.1305E-06	0.001063257	100.00%

USL	0.009
LSL	0
Precision to Tolerance Ratio	0.49786012
Precision to Total Ratio	0.70236076
Resolution	1.4

Histogram

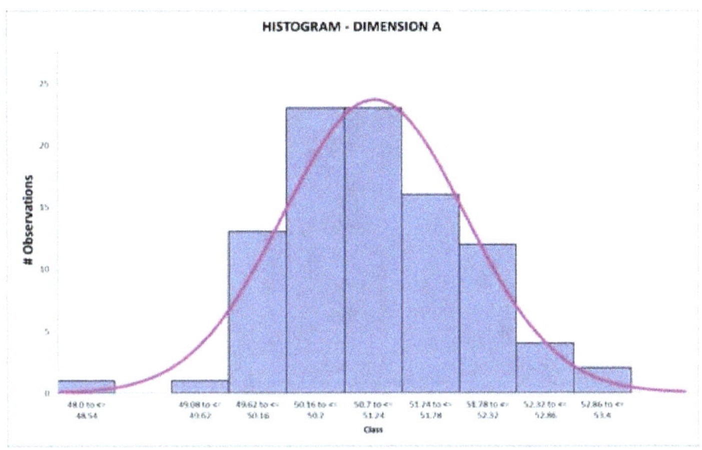

Pareto

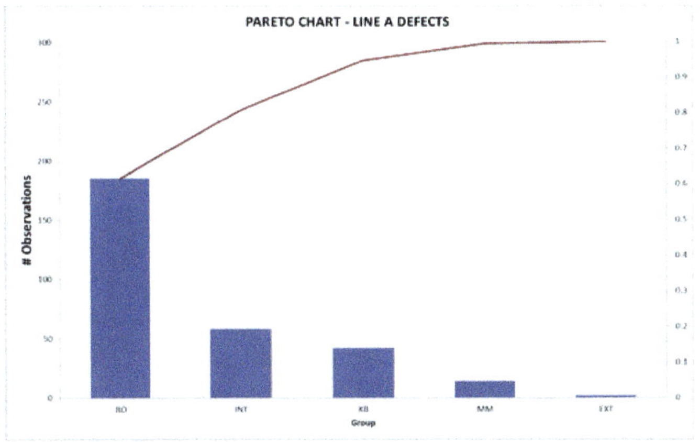

Multi – Vari Chart

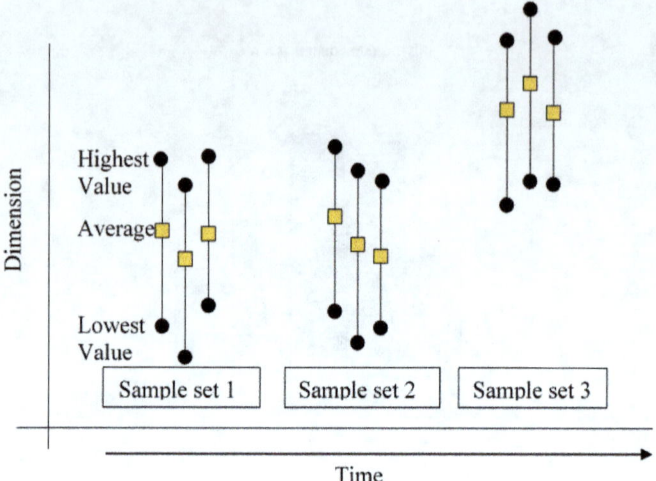

Flow Diagram and the Cause and Effect Diagram

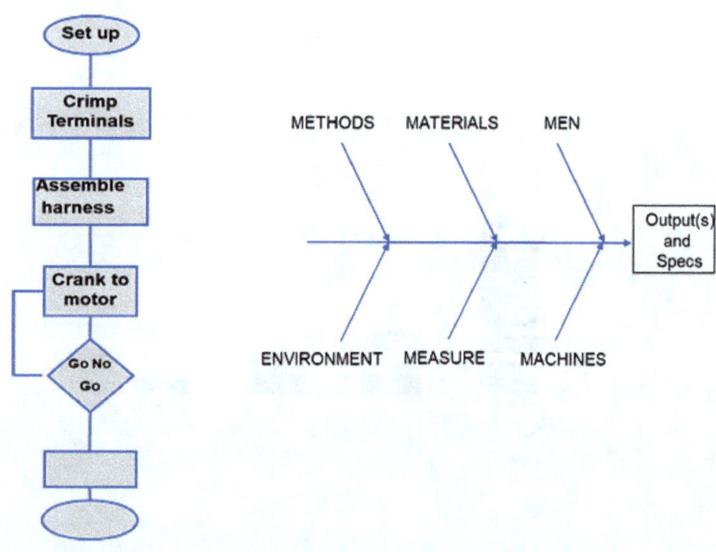

Cpk Diagram

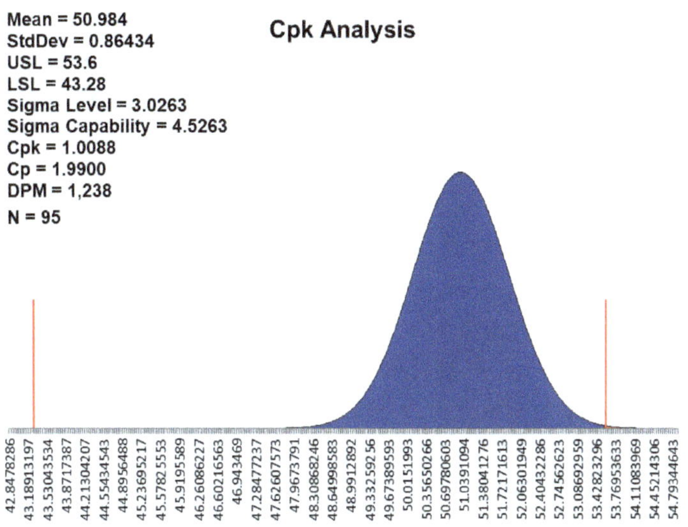

Control Chart/Statistical Process Control (SPC)

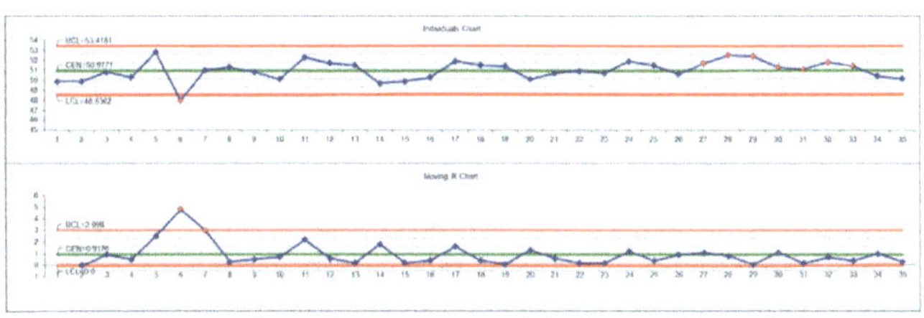

You might think that we will take each tool in turn and there will be a chapter for each one describing a tool and its use. But this would really misunderstand how these tools are used. The tools should always be used together and in order for these tools work correctly and to solve your technical problems you will need to understand some simple principles and how they relate to your processes and also to apply a structured Continuous Improvement process in your project.

Once you understand the principles, which are like simple laws of physics you realise that solving technical problems is more about the way you think than the tools you use.

With full understanding of these principles, you could apply them and solve your problems without ever using traditional Quality tools. But these quality tools are there to take these laws of physics and bring their full power to bear to solve your problem in an easy and practical way.

That being said we are going to look at your Yellowbelt training in 3 stages:

1. Using the laws of Physics to introduce Process thinking principles
2. Introducing the D.M.A.I.C problem solving process
3. Discuss the individual tools

We feel that covered in this order the use of the tools will make perfect sense and the final detailed discussion of the tools rather than being a little dry and pointless will have real practical meaning...

As we mentioned at the beginning, we will also be adding what might be considered additional tools. We will look at simple summary statistics from your process data. This is part of really understanding your process problem and also discuss how to use tools like 5s/Workplace Organisation, ISO9001, Preventative maintenance and Standard Operating Procedures in your problem solving to lock in process control.

In this way not only do we introduce the basic Six Sigma tools and an improvement process to you, but we show how they make a real improvement, linked with techniques that you often already have in your business but are currently being poorly used.

Remember this text is supported by my video's on youtube...

Here is the link type it in now and take a look...

https://www.youtube.com/channel/UCYytJvrE9Ivibz9qTVmow2A

Chapter 1

The Approach

Think about a set of skills that will enable you to walk into any process situation in any industry regardless of your training or background and with the help of the local team, understand and control that process.

These skills will help you become an engineer or problem solver who can control any technical process for anyone! As a Yellowbelt you'll be part of a team using these skills. As part of applying those skills the tools covered in this book will be the ones most commonly used and that you'll come back to again and again.

Explanation of 7 Quality Tools – https://youtu.be/_W2gPOAbZsk

How the 7 Quality tools implicitly use Statistical principles - https://youtu.be/uy5ZVLj56x4

Here are 3 case studies - https://youtu.be/kEfVq7My_Hg

In the video's you'll see an introduction to powerful but simple techniques and how they can be successful on completely different problems and processes. It's good to look at these video's now and return to them later in the book when all has been fully explained.

Over the next 180 pages or so we'll introduce some important principles and then walk you through a process applying these principles using the simple quality tools, we'll then finish off with the dry description of the individual tools and add some case studies in the text at the end to re-enforce the practical use of these tools to remove problems and frustrations in your work place.

In order to help you understand how these different aspects work together we've prepared a flow diagram below. We'll refer to it several times in the text coming back to it again and again to guide through a great understanding and cement your new skills.

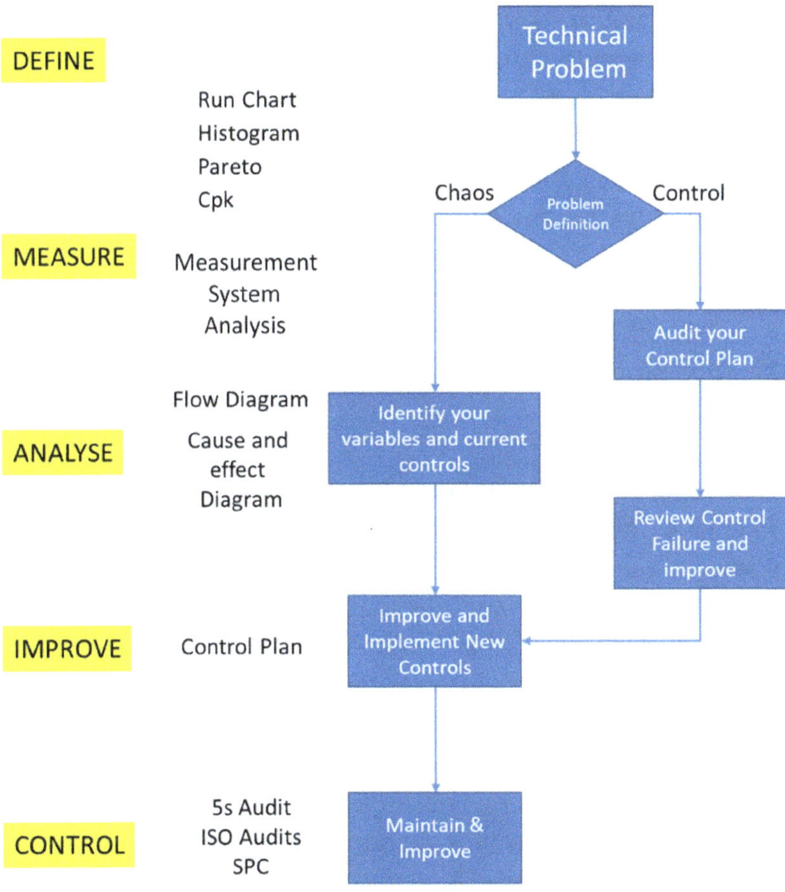

Let the Software Do the hard work

As I mentioned earlier, we're not going to cover the actual mechanics of completing the various diagrams or summary statistics based on your process data

With the common availability of Laptops in most business situations and sensibly priced Quality Software we can cut past the maths and go straight for what these tools are really good at.

What are they telling me about my process physics and how can I use them to make my process perform better than ever.

So as the number crunching is a couple of clicks away **the good practical use of these tools is everything, not manual number crunching**.

I have chosen to use some great software that was specifically designed for quality problem solving to base my examples on. It's called **SPC XL**. It can be downloaded from www.sigmazone.com and has a 10 trial licence if you want to copy any of the analysis in this book. We can then concentrate on how to use the tools and get the most from them.

The software is excel based and loads into the excel ribbon as an extra menu choice. Most business data is already in excel so everyone will feel comfortable using this software. The software is designed specifically for Quality Improvement work and is very easy to use. If you're offered Minitab I would avoid it! Too expensive, too complex, just makes it harder not easier....

Chapter 2

The Physics of your Process

The Laws of Physics and Process Improvement

Teaching you some tools without the principles behind their use is like giving a mechanic a set of spanners but not telling them how the engine works.

Apart from covering the principles we'll also give you a process to follow as you solve your problem. The combined use of the **Principles, Process and Tools** is what will ultimately make you a successful confident problem solver.

Lots of courses just teach you tools and that creates these problems....

- You're filling out FMEA forms, but you don't know why...
- You're filling out SPC charts, but you don't know why....
- You're doing 5s and using it just to clean up...
- You do ISO audits, but you don't know why, other than it puts a certificate on the wall that customers seem to like!!

If used like this, all of these things become additional work, but add no value and get in the way of making money. When they should be tools that enable and help you to be more successful.

It is most likely that you have purchased this book because you already carry out some form of process improvement work. When you take on an improvement project what are the principles that you employ?

What repeatable approach or process do you use currently that could be taught to anyone? Take 5 mins to write down your current approach and write next to it what you consider to be your personal success rate of problem solving.

What does it look like? And how successful are you? And does the person next to you have the same approach? If you don't have a common approach then when you get in a meeting with your colleagues to discuss how to solve a problem, you'll probably spend most of your time arguing about what to do.......

This where the **Principles** and **Process** that you'll be using will collect everyone's process knowledge and push everyone in a common direction.

Firstly, the **principles**. You can see from the flow below diagram that we introduced earlier. The first question we need answer at the Problem definition step;

Is your process in a state of chaos or control?

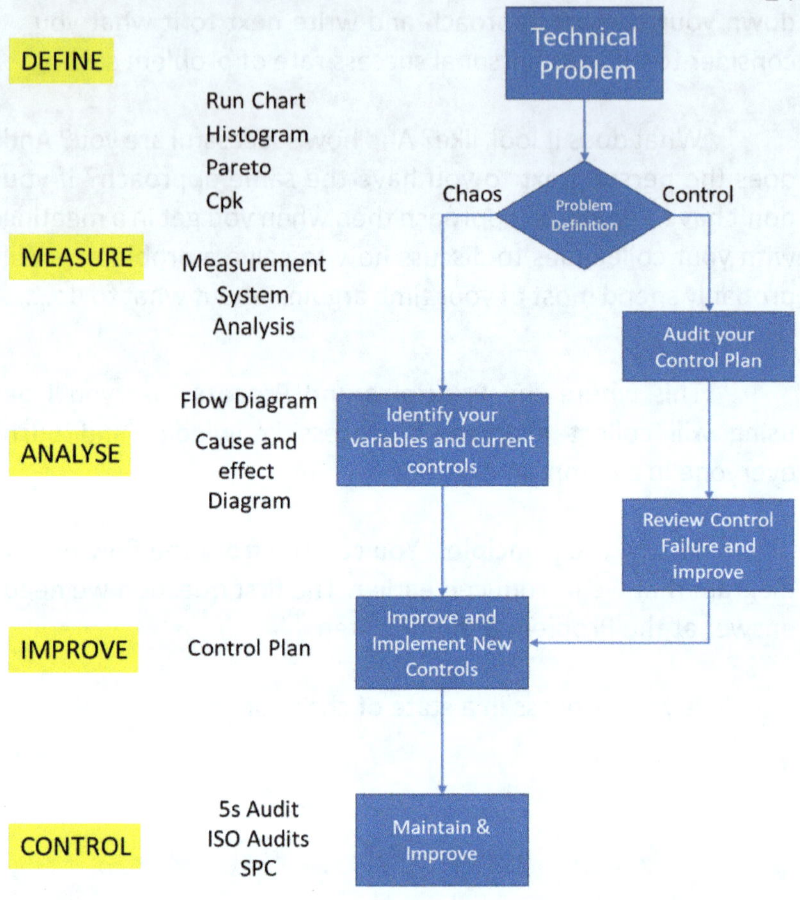

A process that is in Chaos is completely unpredictable and unstable.

To answer this question, we're going to use our first of the quality tools, the run chart and we might add the histogram to this analysis.

Simply take 30 data points and plot the data in order on a line graph to complete a run chart. Do this by hand or use standard excel line graph, no special software needed.

If your process is in a state of **chaos** and most are! Then it will take 3 months to fix the problem. It will need a project and a team to improve this process. This is a process improvement project.....and you're going to need some statistical and analytical tools and time with a cross functional team. An ideal target for the quality tools in this book.

If your process is in **Control** or in controlled Excellence and then a problem crops up. You should be able to fix the problem in 3 minutes, 3 hours or 3 days. One person can usually fix it and the tools are simple process audits to find a root cause and common sense will do. What you're doing is Auditing the current controls you have and one them will be the problem, a rule that is now not being followed, this is the root cause. Fix it, return it to standard and look for the deep reason for this single failure and improve your controls if necessary. This is easy to problem solve.

A Problem of **Chaos** takes 3 months to fix, A problem when a process is in control it takes 3 days maximum and only one person is needed to fix it.

So how do you decide.....**CHAOS** or **CONTROL**?

Well this where tool 1, a run chart, is an absolute necessity. The simplest chart but a chart simply not used enough....

The run chart below is from one of my clients, who build cabins to fit out long haul planes. It measures the amount of rework hours needed to produce each item. Look at the period of chaos. Random amounts of re-work, bouncing between 1 hour and 6 hours. The graph shows rework hours for each job, something that they wanted to eliminate completely if possible. The initial results are totally unpredictable. The graph shows 3 months' work, to get the process under control and now we have 1 hour of rework or less, stable, regular and predictable. This is a process now under control.

Primary Measure

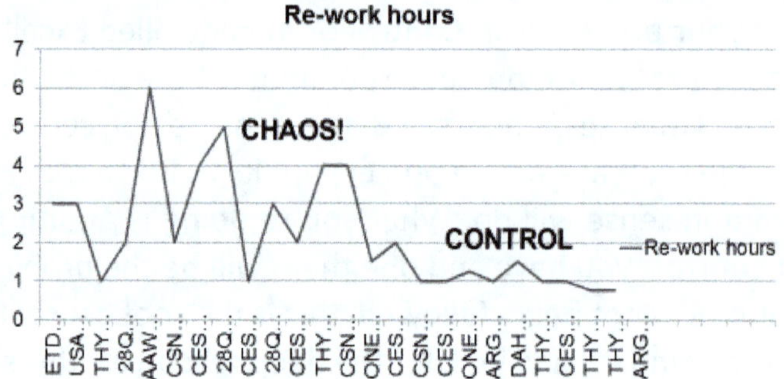

ALLEN & PARTNERS LIMITED

Project Selection [3]

Only processes in **CHAOS** should be considered as a team based 3 month project. If the process is in Control and something goes wrong. See the Graph below. This can be fixed in less than 3 days. It's not a long project. A simple process audit of your controls and then 5 Y's and root cause analysis will be the answer.

This graph below shows a long period of stable, consistent behaviour and then something special happens. The problem gets corrected quickly and the process goes back to the original pattern.

.

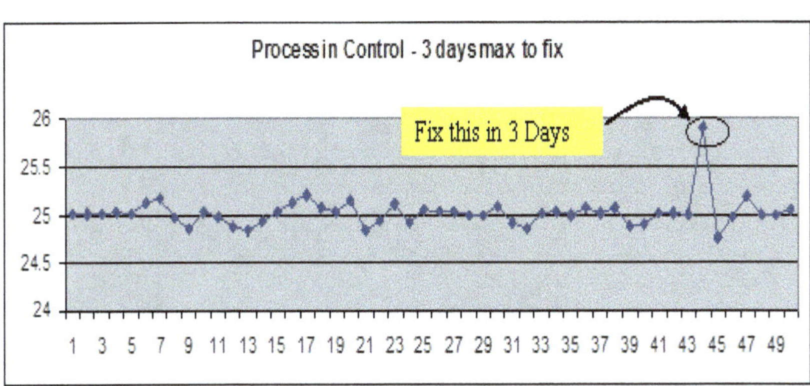

Once you've found a process in **Chaos**, then we can get on with a project and use some of the other tools.

To re-enforce your decision that you process problem is chaotic in nature, you could also use the histogram...

A histogram is a specialised type of bar chart, individual data points are grouped together in classes, so that you can get an idea of how frequently data in each part of the scale occurs. The strength of the histogram is that it provides an easy to read picture of the location and the variation in a data set, across the scale.

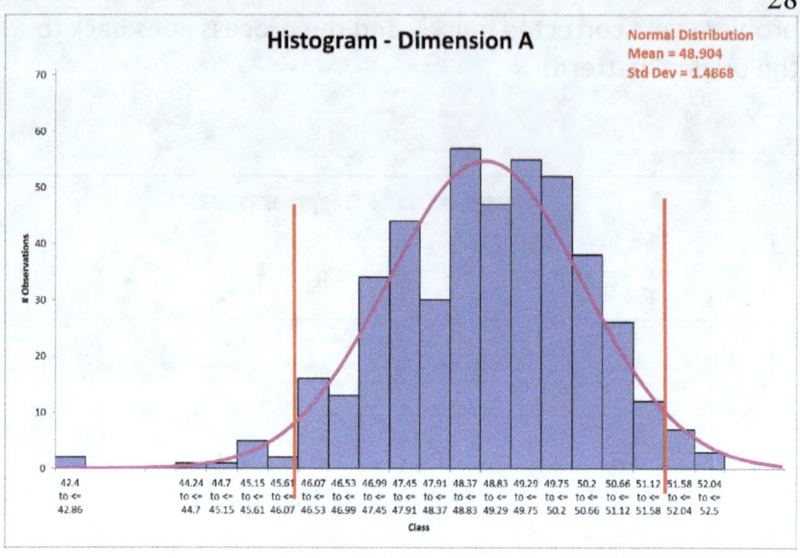

The histogram is a great example of why you always look at data through the lens of a chart rather than looking directly at the numbers. It shows a pattern that you cannot see in the numbers alone. You can see from the chart above that as the data starts to fall in more extreme parts of the scale, down to 44 and up to 52, there are less data points in these areas. More data is falling in the middle of the scale, this creates a distribution that helps us understand the natural behaviour of the process.

On the histogram above, I've also sketched in the specifications we're trying to meet, in the red. These lines wouldn't normally be present, but they illustrate your process problem and can now be added in using SPC XL. The target for this process is 49. The average performance is 48.9, so we're pretty much dead on target. However, without changing settings we're swinging from results that go below the bottom specification to results above the top specification. This is a problem with random variation, another way to describe that, it is a problem with **Chaos** or Noise (The Spread of the data). If the average (mean) was off target, this would be a problem with signal or **control**.

The run chart below shows an initial process situation, a process swinging from too big to too small without any process adjustment. You can see this chaos on a run chart or on Histogram.

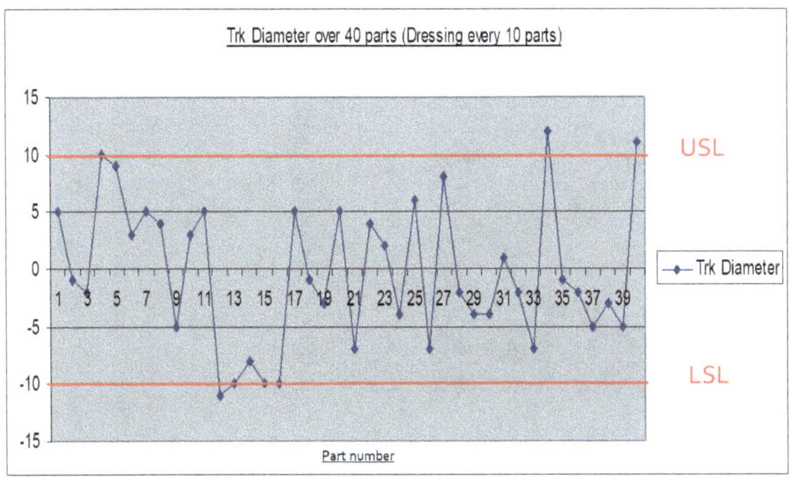

One last point before we move off the identification of **Chaos**, if you have a process that keeps misbehaving with what seems like many one off issues, and you don't think it's possible to get this under control. This is also a process in **Chaos**, although the errors are different each time, it keeps breaking down regularly, it's a chaotic process.

So 1st Principle is my Process in **Chaos** or **control**??

Here's a video to show you why the decision is so important…

https://youtu.be/PKpT7tzjYow

Now that we've identified a process in **Chaos**, we need a set of process thinking principles that can direct us in how to take a process from **Chaos** to **control**.

The Statapult exercise – The demonstration of the 4 process thinking principles.

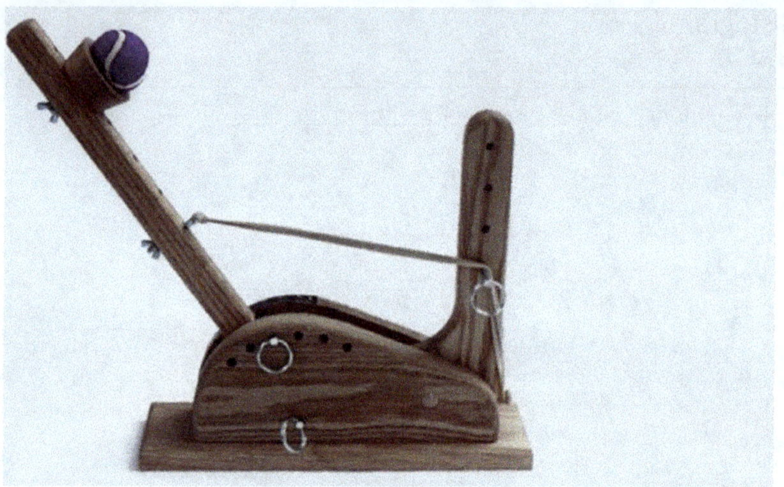

The exercise normally takes the form of a group of people being given a standard set of instructions to fire the catapult. A standard set of instructions that if followed to the precise word would result in the catapult firing the ball an identical distance each time. This would be a process in **control**. Unfortunately, when they try to do this, they find difference in the results up to and beyond 40 inches! 40 inches for the same settings? From this starting point they learn how to really **control** a process and the second time they fire they get a spread of just 5 inches or less, with an ultimate aim later in the course to land the ball in a coffee cup, with the first shot! Yes, perfectly possible.

Here are the instructions for the first part of that exercise.

Exercise: Rapid-Fire Statapult Launching

Objective:

To fire the statapult and record the distance in inches for each of the launches. The measured distance will be from the back of the base of the launcher to the point where the ball hits the floor. Use the following grid to record the distances in the order in which they where obtained (to the nearest inch)

Rules of Engagement:

(1) Every shot will be launched from the pull back angle of 177^{0} , with the peg position as set. Each person on the team will perform the same number of launches. Launching means pulling back and releasing.

(2) There will be a time limit of 15 seconds between firings. Each group will police themselves. There will be penalty associated with late firing. Practice shots and process changes are not allowed.

LAUNCH	ROUND 1	ROUND 2
1		
2		
3		
4		
5		
6		
7		
8		
9		
10		
11		
12		
13		
14		
15		
16		
17		
18		
19		
20		
21		
22		
23		
24		

ALLEN & PARTNERS LIMITED

[28]

You can see me conducting a simple demonstration of this exercise in this video https://www.youtube.com/watch?v=N7BxriI3jpU

Before we look at the principles that you can use to solve any process problem. I want to talk about some behaviour or approaches that are definitely not employed in our problem solving. Behaviours that are very common, you might be using some or all these techniques in your business.

1. We will not blame one thing – In other words we will not look for root cause. This only works when a process was in control and now has a problem. Most aren't in control. This process is in chaos, it has no root cause.

2. We will not adjust the Machine constantly – This is a very bad thing to do. Do you really think that a machine that cost you so much money is so bad that it needs someone to steer it constantly? Does your million-pound machine need adjusting every few cycles. Real process control means you set it, sit back, drink tea read the paper and count the cash!

3. We will not take the machine apart – something the technician loves to do. *'I've got a set of Allen keys and I'm going to use them!'*

4. We won't concentrate on a target either – You can't hit a target until you know where your process is currently set. Look at the graph in the video. Those results were for a constant setting, even if we adjusted and hit the target the process clearly won't stay there, we need consistency first. That is what process control does. You must have control first then look to hit the target.

5. We won't concentrate on output graphs – The output tells us we have a problem and that it's **Chaos**. But once we know that, we'll concentrate on the process that

produced those results. If you concentrate on the process outputs, the only choices you have are to inspect and grade the rubbish out or rework the products that are made.

Ok if they are the 'improvement' methods we're not going to use. What are the methods we are going to use?

Here is the exercise I want you to complete. Look at the video of me firing the catapult, you can see me collect data on my 24 shots and plot it simply on a graph. The Graph shows that we have a chaotic process.

Watch the video again of me firing the catapult and complete a Flow Diagram of the process you can see me go through….

Load Ball……Pull Arm back…..check angle…..etc etc.

Now use the flow diagram to complete a basic Cause & effect Diagram like the one below. As you do this there are some very simple but very important rules you must use to complete the cause and effect diagram…

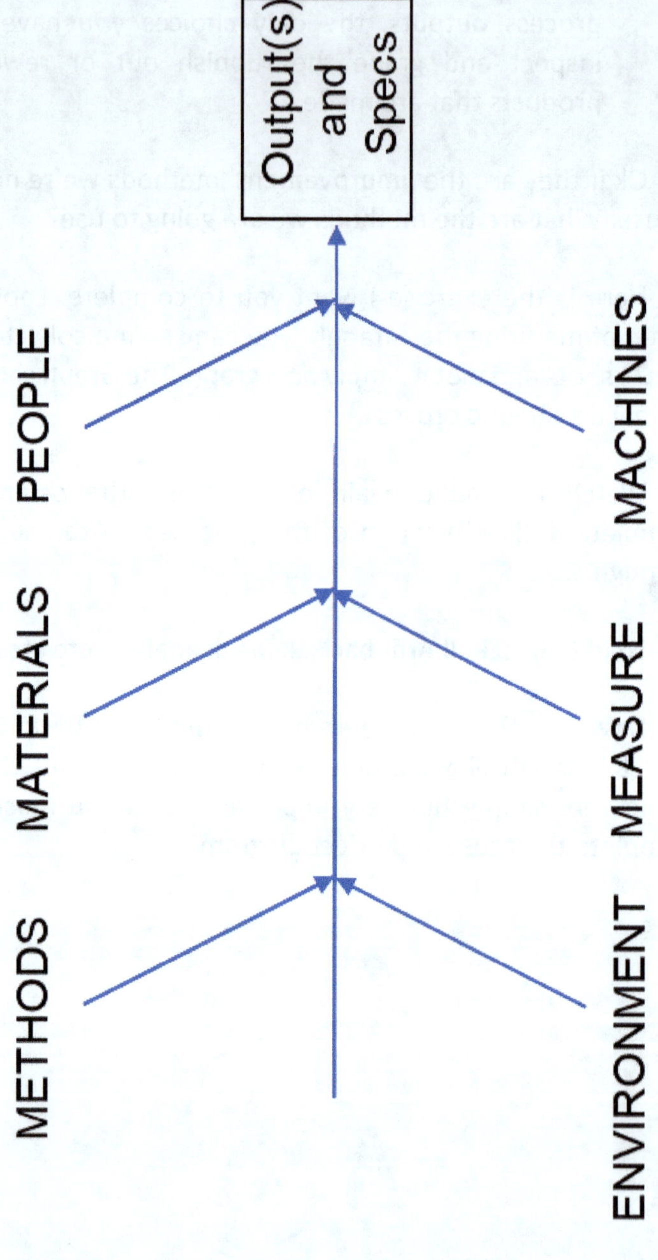
Output(s)
and
Specs
PEOPLE
MATERIALS
METHODS
MACHINES
MEASURE
ENVIRONMENT

Go to each box on your flow Diagram and ask this simple question.

At this stage of the Process what are the variables that contributes to the 43 inch range??

As you discover a variable simply write it on an appropriate leg of the cause and effect. The key point here is that we are not asking you to identify problems. We are asking for anything that I'm doing in the catapult process that is different or changing each time I fire it.

You can see me go through this part of the analysis process in the video later in this section. Complete your analysis first then watch the video to confirm your learning.

Once you've identified all the variables, I want you to list all these variables and then identify a control that could reduce or eliminate the variability. This is what process control means, removing variability. In a classroom setting the groups that run this exercise are given several tools to help them with this Improve/control phase:

1. They could write a good specific Standard Operating procedure.
2. They can use Kitchen foil
3. They can use packing tape
4. They can use talcum powder
5. They can use Masking Tape
6. They can use any other sensible item in their pocket or in the room

This idea of writing down an input variable and then designing a control to eliminate as much input variability as possible is a true control plan.

I've created the flow diagram and a control plan solution template on the next page, before you look at it complete your own process flow cause and effect exercise and complete a control plan now.

Here is the video showing the complete Process Flow, Cause & Effect Analysis. With individual controls designed and control audits linked for longer term control.

https://youtu.be/Sf_DT_pkMCw

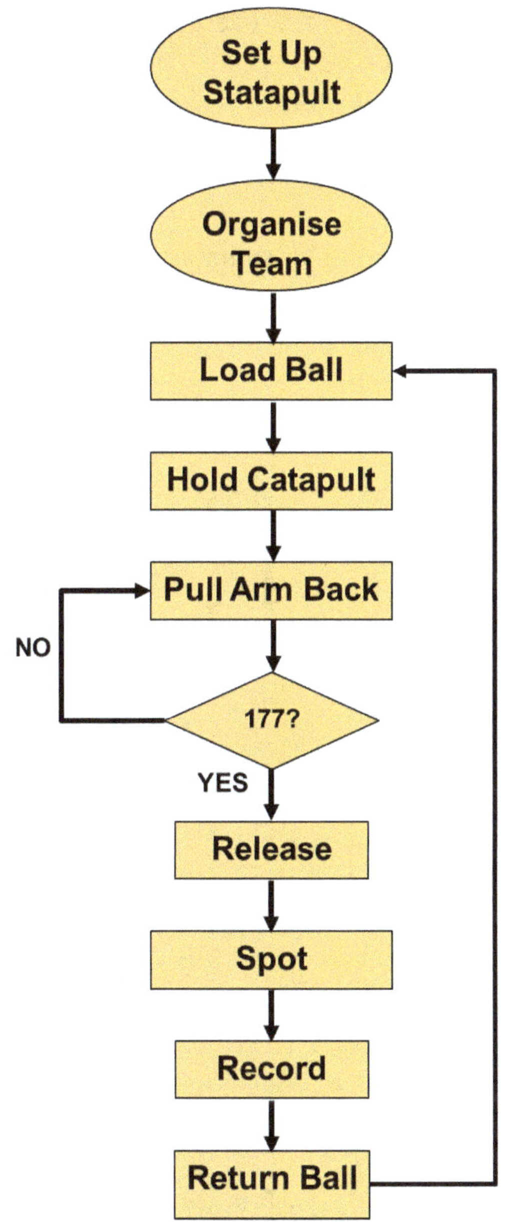
Set Up Statapult
Organise Team
Load Ball
Hold Catapult
Pull Arm Back
NO
177?
YES
Release
Spot
Record
Return Ball

This is a process you'll need to repeat in solving your process control problems. Using the flow diagram above we've identified the list of variables below and then designed controls and added an audit process that keeps these implemented controls in place.

This is the summarised table, firstly we would have completed the Cause and effect and then listed them and designed the controls to give us this table.

Variable Controls Planning

Variable	Control Method	Audit process
Stability of Catapult	Tape Down	Maintenance
Stability of Tape	Tape Down	Maintenance
Position of Operators	Mark out workplace	5s Audit
Training	Clear unambiguous SOP's	Audit
Condition of Elastic	TPM	Maintenance
Shape of ball	Alignment Marks/SOP	5s Audit
Load Force	SOP at the Point of Activity	Maintenance
Steady Method	SOP at the Point of Activity	5s Audit
Arm Hold Method	SOP at the Point of Activity	5s Audit
177?	Mistake Proofing Back Stop	5s Audit
Pull Back hold time	SOP at the Point of Activity	5s Audit
Release method	SOP at the Point of Activity	5s Audit
Spot Ball	Foil in landing Area	Maintenance
Measure Method	SOP at the Point of Activity	Audit
Return Method	SOP at the Point of Activity	Audit

ALLEN & PARTNERS LIMITED

6 Sigma 1 Day Awareness [12]

In the classroom workshop we now implement these controls and then fire the catapult again. This time producing a controlled process result that lands within a 4-inch range.

In the video the last 24 shots are plotted on the same graph as before... Now we have a range of 5 inches and a process under control.

Now it may not have been obvious but during this exercise, you've used 4 of the most important process principles you'll ever learn. Principles that you'll now use on every process problem you work on, with or without these quality tools.

The idea of this exercise is to introduce these 4 important process control principles and to show anyone how to use Process Flow and cause and Effect tools to deploy these principles

What are the 4 simple principles that anyone can use to make this level of process control transformation?

1. **Control inputs not outputs** – Have you ever heard that well used phrase about computers, 'rubbish in, rubbish out'. Well if you really think it through it doesn't just apply to computers, it's a basic law of physics. If you want to make a process do what you want it to do every time you use it. You need to control the inputs not the outputs. In the catapult exercise I asked you to use the flow diagram to find variables and put them on the cause and effect diagram, but another phrase I could have used is describe inputs that effect your problem. What we did was identify inputs and then make sure that all the inputs were as consistent as possible, by

controlling the consistency of the Inputs we guaranteed the consistency of the output. This is process control. This is a principle I use every day and makes ever process problem easy to fix. Identify inputs and fix them.

2. **Variation in = variation out** – lets go back to that catapult. Imagine you are trying to hit a target defined by a customer. But every time you set it up and fire it, it doesn't land in the same spot instead you get a different distance up to 30 inches different, shortest to longest as we had in the video. How the hell can you hit a customer target when you can't hit the same place 2 shots running, with identical settings! If you let variability into the inputs, then it simply adds up to make the output highly variable. Variation will not cancel itself out. It just multiplies and gets worse and worse. So if your suppliers give you variable raw materials, you cannot adjust this out by changing the machine settings. That way you'll only add to the variability of the material and make it worse. Most companies deal with this by inspecting out the defects, a very expensive solution. This is a very important principle and goes against human nature. Fix the settings on your machine, take your hands off, drink tea and read the paper! The machine will run so much better...

Don't believe me? Well let's look at a simple process.

A dice.

Now we know the dice will vary between 1 and 6 naturally. let's assume though we don't like that and we want it to land on only 3, 4, or 5...inside 3-5 is our specification.

As we have a chance of making defects I'll give you an offset dial, that you can adjust after one defective result, to make the process 'behave' for the next result. You are now in control of the result and this is what technicians do a lot, adjust after a bad result to try to make the next one better, let's see if you can make the process land on 3,4.or 5? You'll have a process that looks like the IPO (Input process output) diagram below.

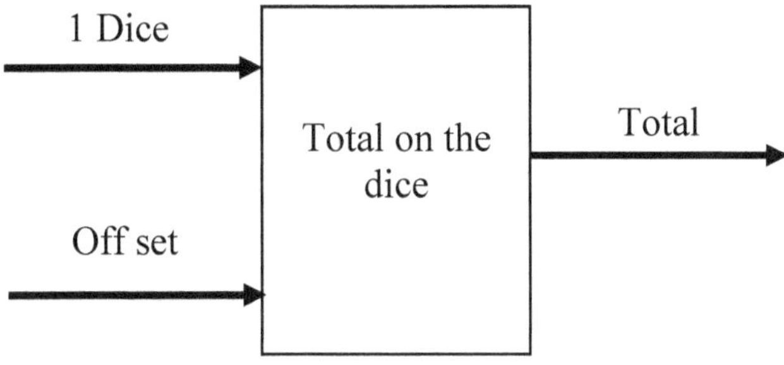

Now before we start controlling the process and making it better what is the potential max and min? 1 and 6.

Obviously if you get an answer 3,4 or 5 you won't make an off set.

Now what will happen if I get a result of 1? I set the off-set to +2.

This doesn't affect the current result; you've already produced that one. This would make the minimum of the next result 3, inside our tolerance band. But what if the next result on the dice is 6 ? With a +2 offset suddenly I can hit 8! at the top end and when I do hit 8 what will be your response? turn the off set the other way of course! Maybe to -3...now I can hit -2 as a lowest result....so before I gave control to you, we had a total variability of 1 to 6 now that you're making the process 'better' with the off set, we have a variability of -2 to 8!!

We seem to have made it worse!

And that's what operators and technicians are doing most of the time. Making the process more variable. Getting control is not about adjustment. It's about fixing as many inputs as you can. Then take your hands off, drink tea and read the paper!!

3. **Eliminate Variation 1st, hit targets 2nd** - Most companies chase the target. They try to chase the average result. The average result is known as the signal. But moving the signal is not important when you have lots of variation in the process. What you should chase is the variability, the noise in the process. So for the same settings you get a similar result, If you make the process hit the same place every time, hitting the target will be easy if you adjust one of your settings to a better place. However, if the aim of your process improvement is not to hit a specific target but just to drive the results higher the better or lower the better, even then you should tackle variability first. You see, it's very difficult to be consistently crap at something! If you get the process

consistent it will naturally be consistently good. So if you remove variability 1st, you'll either hit the target, because you can easily adjust to target or you'll hit the target naturally because consistency drives excellence. Look at your graphs, do they jump around, up and down. Get rid of that variation!

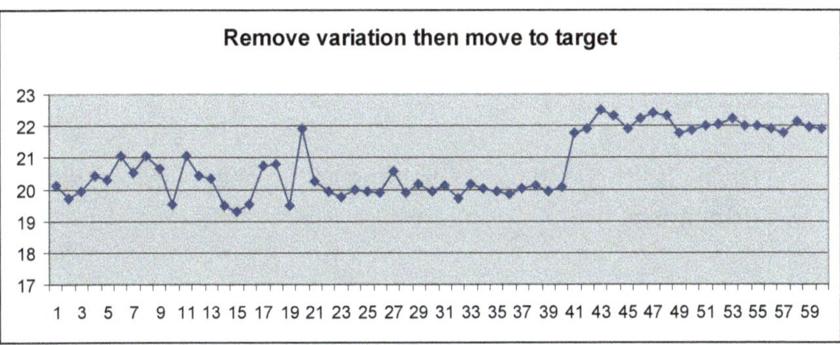

4. **Work on the process not the product** – Do you ever find yourself attending an inquest? Everyone is sitting around inspecting the rejects or customer complaints trying to figure out what caused them. They are often talking about finding 'root cause'. These rejects and events are useful to direct your improvement activity but you are looking at outputs. You are at the wrong end of the process. At this point, there is only one thing you can do, inspect & rework or remake. Data analysis is useful but only by transforming the process will you transform the data. The tools we will look at later will get you to concentrate on the process. This is a hard principle to apply though. After all the customer is very interested in the product, therefore you spend all your time trying to get the hot order out the door and once its gone you relax...... Phew! 'Thanks god that's over' and you forget to go back to fix the process that created the

mess in the first place and of course at some point the problem happens again.

These principles should be clear in your mind whenever you tackle a process improvement project. However, if you want to sum it up in one simple phrase it would be:

Process Improvement & Control in a Nut shell – THE RELENTLESS AND RUTHLESS PURSUIT OF VARIATION REDUCTION!

Here is the video explaining 4 process improvement Principles:

https://youtu.be/1vpCbEBr2o4

The Cost of Variation

Random variation affects your ability to improve and make good decisions and it also blinds you to real process knowledge.

Imagine you have been called out on to a process to deal with a quality problem. If you look at the graph on page 41. When would production ask for help? On the worst ever day of course, when the graph peaks. Therefore, even if the action you take makes no real difference at all what do you think is going to happen to the results next day anyway? Well you have un-controlled variation, so the result will definitely be different and obviously better than yesterday, since yesterday was an extreme day probably a lot better. As a former Industrial Engineer this is a scenario I have been through hundreds of times. Called to a process to help I would fiddle about with something. Of course I would link that 'improvement' to my actions. So I would sit back in the office with a satisfied smile on my face and pour myself an extra cup of coffee, I've earned it! When do I get the second call to help? At the next extreme result of course, I repeat my actions and the results improve a second time. Now I'm a process expert!

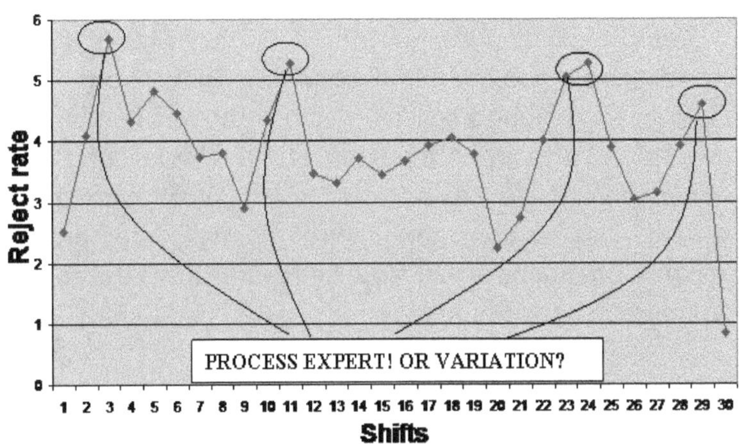

At the same time another Engineer might also be doing the same thing on a similar machine. But on a different section. He might be taking different action to me, but seeing the same type of 'Improvement'. When we discuss these processes, we have differing view of what is important and what to change to make the process improve? Have you ever sat 3 technicians down from 3 different shifts and wondered why they all have different views on how a process works? Variation!

Actually, I'm not picking on process engineers here, you can pick any theory you want that you think is the 'root cause' and can prove it to be 'true' when extreme variation is present.

Let's think of a senior Manufacturing Manager who notices that today has been really bad day. They go out to the shop and give everyone a rocket, next day the result improves! He now starts to think all this stuff about looking after your employees is all nonsense and when a bad day happens again, he repeats his actions and sure enough the process improvement happens again. Now he is sure this is the correct way to improve performance!

How about raw materials? Everyone always complains about the raw material. You change the batch of material, check the results next day and.....yes, you've guessed it, the process improved!

Now if you sit this group of people down who have worked on 'improving' this process and ask how do we get better performance? They say....

Adjust the machine!
Kick the staff regularly!
Change the supplier!

And all these people have a graph that proves they are correct!

Of course, if you look at the problem correctly, the graph is not showing any improvement just randomly bouncing up and down as processes do. It's extreme variation.....

- Variation is affecting your decision making and process knowledge – Get rid of it!

And if you use the 4 principles getting rid of it, is actually very easy...

- Inputs control Outputs
- Variation in = Variation Out
- Variation 1st Targets 2nd
- Work on the process not the product.

And the great thing is once you use these principles every process looks the same. A series of inputs and a number of outputs. Just like the diagram below. This is just process thinking.

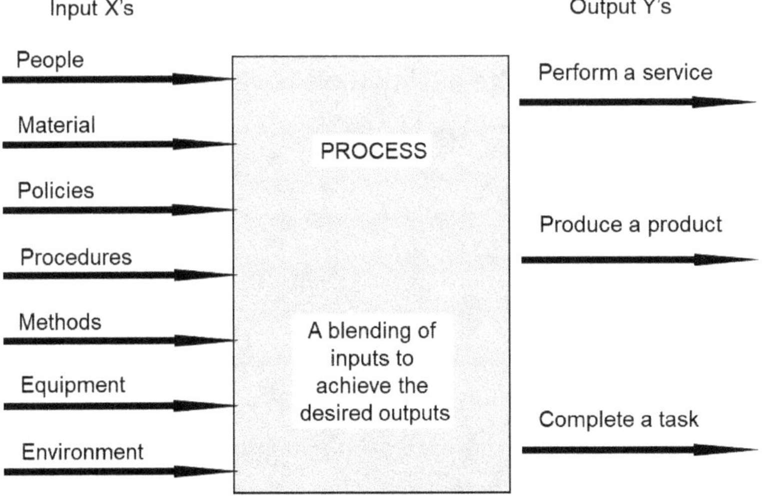

If you fix the inputs and therefore remove or control input variability. The laws of physics say you will get outputs that are consistent. And the more consistent the outputs the easier it will be to hit any target your business or customer requires.

Think about this diagram for a second. It contains one of the most post powerful principles you'll ever learn in process improvement. Every process is the same!

And most project solutions are the same, fix the inputs!.....it doesn't really matter where you fix them, just fix them and see what happens.....and that's it!

If a variable doesn't move it can't hurt you. It can't be a problem. And if it's not the problem, what are you not going to do. Change it!

Go and find the real problem and put it back on standard.

Now that we know that variation is the enemy of process performance the quality tools have context and we'll know how to use them...

Chapter 3

The Improvement Process

Now we know the enemy of process performance, let's take a look at the process that we are going to use to improve it.

A little earlier I asked you to write down your improvement process, what does it look like? Does it require you to have prior process knowledge? If that is the case could you take it to an environment where you have no knowledge and make it work? If the answer is no, then it's not as useful as it could be.

The ability to be called into a new environment with little process knowledge and help the team solve a long-standing problem would be a pretty useful skill. It is very much what consultants who work in the quality field do and to let you into a little secret the way this is achieved is by using general, repeatable principles and a repeatable improvement process. Plan, Do, check, Act (PDCA) is well known example or indeed the 8D process...

The Improvement process we've decided to use is: DMAIC

Define
Measure
Analyse
Improve
Control

Now it's not rocket science but these 5 simple steps are most important if you really want to improve the process, know you've improved it and make it stick. If you really want to fix a problem for good, then you need to become good in all 5 of these phases. The 8D problem solving process that some businesses use also contains these 5 phases, but often it is not explain thoroughly how to carry out each step successfully.

As a production engineer in a manufacturing environment my basic approach was to go straight to I-Improve. I am a very helpful kind of a bloke. If I received a call from production saying they had a problem, I would get straight out into the shop take a look at a few parts, tell them to re-set control knob A and re-set control knob B. Take a look at a few more parts, see that the problem had 'improved' and stroll away thinking I'd done a good job.

Of course, a week or so later I would get a call to say the problem had returned. Back out on the process I would do the same. Twiddle a few settings check a few parts and of course good old variation would make sure that I (and everyone else) thought I was improving the problem.

When in fact I had no idea of the scale of the issue. No idea what 'improvements' I'd made and I certainly never left any controls behind. So in reality I never really fixed anything....D.M.A.I.C made a dramatic improvement on my personal problem solving process.....

One last point to make before we look at the individual elements of D.M.A.I.C. This is not a linear set of steps, to suggest that it is linear ignores the basic practical approach to solving a problem. Don't get me wrong you will start at Define and end at Control. But there is a good chance that you will measure, analyse, improve and control several times before you are happy with the result. Also, there are blurred lines between these elements. Before you can define properly you have to measure for example.

The main thing is we accept that in order to solve a heavy duty technical problem we have to work in the 5 D.M.A.I.C areas and be good at all of them to solve a problem thoroughly and professionally.

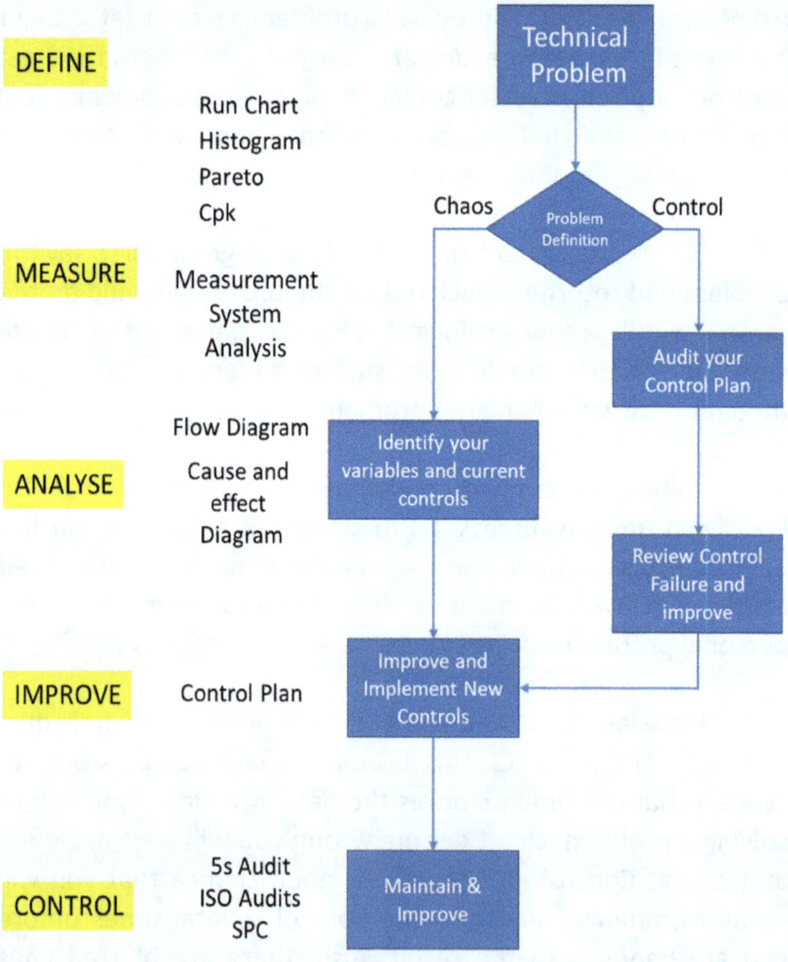

Now that we've outlined the D.M.A.I.C process let's take a look at each stage and position the tools that you would find useful in each of them.

Chapter 4

Define

For me this is the most important phase of the 5 (closely followed by CONTROL). This is because we are going to do more than just define the problem, although that was pretty much how I first used this phase. However, after several years of experience I have tuned this section to define the following.

The problem – Chaos or Control
The Cost of Poor Quality (COPQ)
The target
The team
The primary measure
The data collection process
The Process that is responsible for the problem

The Problem

There is a good chance that in order to do this step you are going to have to carry out some measurement and possibly some data analysis. From this we are going to want a good clear Problem statement and the first thing you need to decide? Will my problem take 3 days or 3 months to fix?

So the first stage of problem solving because it is so fundamental to your problem solving approach is to decide is you process in a state in CHAOS or CONTROL...

This first decision immediately brings in the first of the quality tools. You'll want to use a run chart, a histogram and maybe a Cpk diagram....

You've already seen in the Catapult exercise the use of the run chart to see a chaotic un-controlled pattern. Later in the text we'll look at the histogram and Cpk diagram and how they will also reveal the current state of the process...

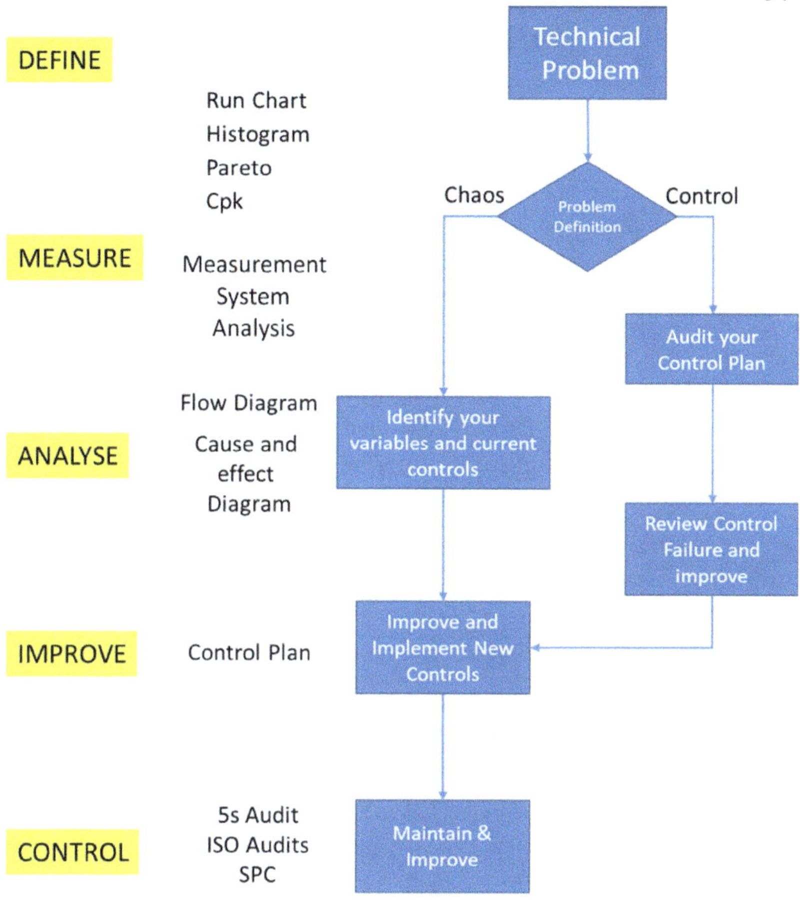

Stage 1 of the problem statement, CHAOS or CONTROL?

Stage 2 how big is the problem? How long has it lasted? How much does it cost?
Now we can turn it into a good problem statement.

A Poor Problem Statement

Product returns are too high and will be reduced by analysing first and second level Pareto Charts

The example above is too vague on the scale of the problem, cost of the problem and too specific on how to fix it. Problem and target statements should never contain phrases about solutions. If the solution really is that obvious then just get on and do it. Don't turn it into a CI project just to tick a box.

Good problem statement

Product returns are 5% of sales, resulting in a business unit negative profit impact of £5M and reduced market share of 10%

Here the Problem statement is more specific. It contains the primary measure (product returns as a proportion of sales). The cost is quantified and other business impacts quantified, if I'm being picky, I would also have included a timeframe. EG 'For the last 6 months' product returns have been 5% of Sales'

So as a checklist the Problem statement should include:

- The primary measure – backed up by data, no guesses
- The time frame of that performance.
- The cost of that performance – but keep it simple
- Other business impacts like overtime to make up for lost production etc.

The target

This is where things can get out of hand. Don't' set too ambitious a target. That makes no sense what so ever.

Here's a simple way I do it...

Take a look at your primary measure on a run chart identify the best results, probably only 3 or 4 points on the chart. Here is a realistic result that your process is capable of, if we could make your process perform here every day instead of 3 out of 30 days we would halve the reject rate! So halve the reject rate would be a good target for this problem, one that is likely to be achieved. Don't forget targets are always a bit arbitrary and never have a link to the actual result that your solutions will deliver, because you don't know what your solutions are at this point in time and really how capable your machinery is at producing good product.

Experience has shown me though most projects can outperform the 'halve reject rate' target.....that lots of people use.

A Poor Target Statement

Reduce product returns by implementing individual performance measures and objectives.

Again this is too vague and has solutions in it.

A Good Target Statement

Reduce product returns of product line xyz from 5% to 2.5% of sales by year end 2013 to reduce overall business unit returns by 1%

There is a nice acronym that helps get the target right SMART

Specific
Measurable
Achievable
Realistic
Time bound

The target in the example is based on your primary measure, quantified and time bound so it covers the main SMART elements of course the additional info that is not mentioned in SMART is the value to your company in terms of money and other benefits. Now this would be another area of caution. Be conservative here, in fact stick to the hard facts. If you are dealing with a reject rate, no one is going to argue if you evaluate the amount of material that is going in the bin.

So a checklist for the Target Statement:

Includes the primary measure
Realistic new level of performance
Values the hard benefits in money
Has a deadline date

The Team

The first point to make about selecting a team is that you have a team. Don't think that because you have worked on this machine/process for 10 years you know all there is to know and don't need anyone else's input. Your team is crucial.

The second key element is that they are mostly doers, operators, technicians, QA personnel, Maintenance technicians, and toolmakers. Supervisors and managers are useful but of secondary importance. Management are really sponsors and give the team permission to work on the project and provide additional resources. It's good to try to way up the skills you will need to make change happen on this process and include those people in the team. At the meeting when we say something is going to happen the chances are the person who is going to do it is in the room. This may also mean you invite a new team member for a short period to complete something specialist then remove them from team meetings when their job is done. Keep the size of the group sensible, 4-6 is about right

Checklist:

- A team
- Mostly doers
- Skills to implement improvement ideas or data collection
- Approximately 4-6 people

The Primary measure and data collection process

The first point to make here, this should only be one measure. I guess that's why I've called it the primary measure. Too many projects are instructed to fix everything, improve the reject rate and increase the output of the machine by 10% and improve the Schedule adherence by 15% for example. It's not that this can't be done, but to use three measures like this is just plain confusing. Simply pick one of them and then get on with improving it. The likelihood is that all the measures will improve at the same time anyway. But if not, start a second project later.

Now apart from picking one primary measure, this part of the define phase seems pretty straightforward. But it's where lots of time and effort can be wasted and your project delayed if simple errors are made. If I was asked were do you typically encounter the first barrier to improvement in most projects it would be at this step, Why? Because most companies have not set up data collection process in order to analyse or use the data effectively. What we should be aiming for is a simple data set that can be used to produce the basic statistical charts in seconds; the primary measure should be shown on a run chart and updated daily or by shift.

Material Interruptions

Date	Part no	Area that stopped	Order Process	Reason Code
10-May	420372351	CTV	RO	1
10-May	Y9000859A	PVM	RO	2
10-May	A1608148A	CTV	RO	2
10-May	121661191	AM	RO	2
10-May	X42004981	CTV	KB	2
10-May	504051401	PVM	RO	3
10-May	166320914	AM	RO	3
10-May	875946719	AM	MM	3
10-May	875958956	AM	RO	3
10-May	113719491	AM	RO	3
10-May	143144221	AM	RO	3
10-May	171158817	AM	RO	4
10-May	177072311	AM	RO	4
10-May	110763691	PVM	RO	4
10-May	420341303	CTV	RO	4
10-May	873505401	CTV	MM	4
11-May	420372341	CTV	RO	1
11-May	154379811	DISP	RO	2
11-May	405051401	CTV	RO	2
11-May	X42004126	CTV	KB	2
11-May	X42004393	CTV	KB	2
11-May	143392511	CTV MT	RO	3
11-May	141255221	AM	RO	3
11-May		DISP	INT	4
11-May	420397601	CTV	RO	4
11-May	150876612	CTV MT	INT	4
11-May	110790391	DISP MT	RO	5

Obviously to get this data you might need to design a data collection process. Maybe only the primary measure is recorded currently, and you want to drill down into failure modes etc.

Once we have the data, we can then start to use our list of quality tools to answer the question chaos or control etc.

However, we are going to save our discussion about the use of your data until we've covered the measurement phase.

In a real practical Problem-solving project, it is likely that you'll carry out the Define and Measurement phases at the same time.

Defining the Process

We've looked at a great tool to define your process already. It's the process flow diagram. Use it at this initial phase of the problem solving project to save time in your Cause & Effect diagram and to ensure you successfully identify as many variables as you can in the analysis day with the team...

Below is a template that you might find useful to enter most of the important information about your project in the define phase..

Six Sigma Project Tracking				
Specialist Name:		Project Title:		
Problem Statement:			Objective:	
Process:		Process owner:	Champion:	
Suppliers:			Customers:	
Tools Used				
Define	Measure	Analyse	Improve	Control

Chapter 5

Measure

OK so we are going to use statistics to try to help us understand and improve our processes, that means collecting some data. Setting up a reliable measure. Setting up the primary measure. We've already discussed setting up your data collection systems.

A key point to make is that,

There should be: No data collection without Analysis
No analysis without action
No action without review

So what sort of data are we going to collect and should we think about it too hard?

Typically, there are two types of data that your process could be producing and that we can collect and analyse. These 2 types of data are:

Variable data - something which measurable, like millimetres and volts any result is possible on a scale. This is often known as continuous data.

Discrete data – essentially counting things like defects, where you can only have whole numbers as results. After all you can't have 1.512 defects can you? So Discrete data is created when you assess products as pass or fail.

Why data types are important to your process improvement work?

Does it matter what type of data we collect? We often set up production equipment to simply give a pass/fail result, a red or green light or use a plug gauge go/no go to simplify things for the operator and that is certainly a good thing.

But if the jig has to take a measurement in order to make that judgement should you keep the results, the actual value or is pass/fail information good enough?

This is from the point of problem solving, not production speed.

Let's consider 2 sets of data from processes producing identical defect rates.

Process A results and Process B results. Both of them are causing us problems with a high defect rate and we need to get stuck in and fix these problem processes.

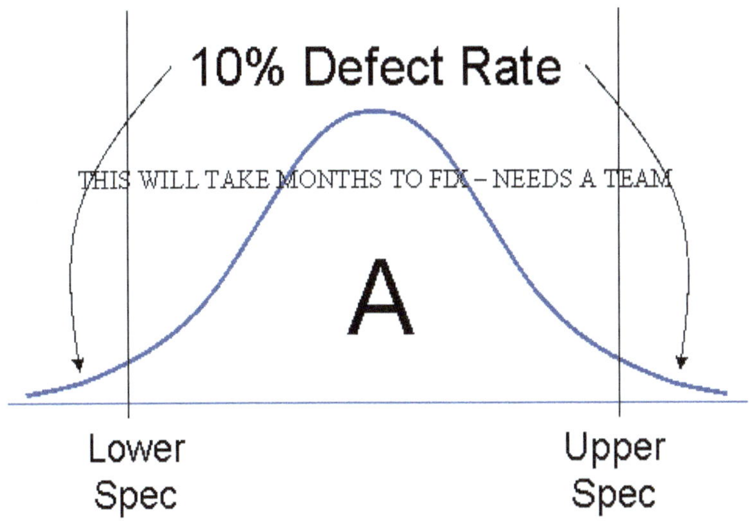

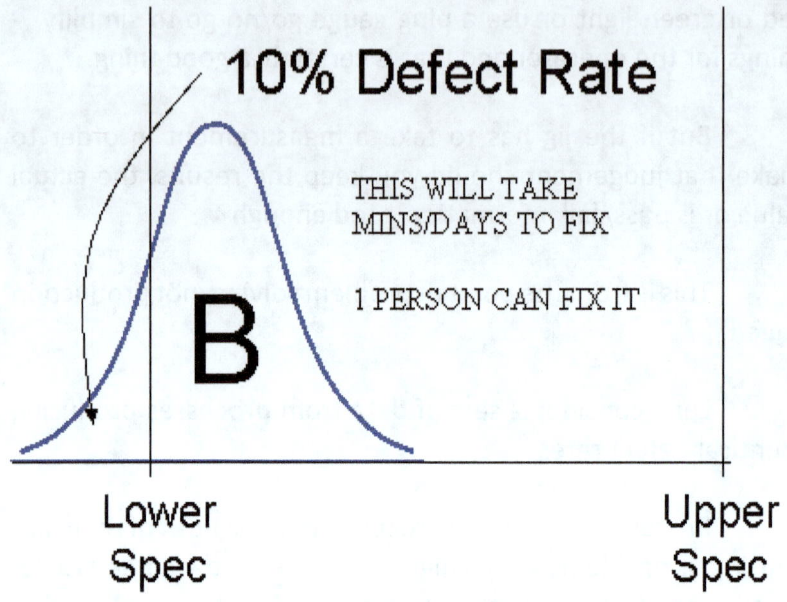

Process A has a 10% defect rate, process B has a 10% defect rate, the same. But are these problems the same? Solving those 2 defect rates is completely different. Process A must have variability removed, it is essentially a process in Chaos, process B simply needs to be centred. To remove the variability from Process A it will typically take months to achieve, it will need a team and you'll be using the Process Flow and Cause and Effect technique we've already looked at. Senior Management will need to encourage and support this type of work to take place.

To centre Process B will be a much quicker fix. Possibly minutes, days, certainly no longer than weeks to fix, this is often one change. Changing a single setting or repairing something that has failed. This can usually be the decision of one person at the point of activity. To know what type of problem you are trying to solve is very important.

Pass/fail discrete data would simply view process A and B as the same they both have a 10% defect rate...

This extra information can only be obtained from variable data; it is data, which can be described as information rich. Pass/Fail data, on the other hand, is information poor and as such requires you to collect much more data to obtain the same level of confidence in the estimate.

Typically, variable data is happy with smaller sample sizes of 30 to 50 in order to make sure your estimate is as good as it can be. So if you had a box with 1 million parts in it and you wanted to know the defect rate, then with just 30 pieces you could estimate very closely what the defect rate is for all 1 million pieces. Again, that's the power of using statistics. Prediction.....and by the way it will make this prediction even if there is zero defects in the 30 samples. The Cpk diagram will do this.

Whilst discrete data requires sample sizes of 1000-3000 and even then, the accuracy of the pass/fail estimate will only ever be +/- 1.5% at best.

Remember those two rules of thumb. As they will be very useful when you have results in your project. We often get distracted by a small sample size that is defect free and we think we have fixed a problem.

Variable data – good sample size 30 to 50
Pass/Fail or discrete data – 1000 to 3000

As process improvement experts you always want to see the variable data. Make sure it's always available, even if the jig has a pass/fail lights on it. The measured numbers must be made available. Red/Green lights are OK for production, not for process improvement investigations though...

Later on we will discuss some common chart techniques and when discrete, information poor, data is used sample size is going to be very important to the way the chart is used.

One last point, of course it could be that your problem is cosmetic defects and that there is no way to get variable data. That is ok you can still proceed, but if you have a choice get variable data to help you...

Plot the data! Plot the data! Plot the data!

Now we are collecting the right type of data we need to turn it into information, So.......

Plot the data!
plot the data!
plot the data!..........

This is not being said for emphasis, but you really should plot the data 3 ways using 3 different charts. Each chart will potentially use a different lens on the data and therefore show a different trend or signal.

In our list of quality tools we have several diagrams that we can use to do this.

Run Chart
Histogram
Cpk
Pareto
Multi-vari Chart etc

Certainly a good piece of advice is to never ever look at the numbers on their own, they are typically useless to you. The main use for the data points is to use them to draw pictures. Too many weekly management reviews simply look at this week's single data point. If the result has improved they have a party! If it's gone down, they have a hanging! This is pointless, look at a chart and get more information that covers 3 months of performance and plot in your charts. Avoid this type of behaviour and look at the variability in the chart!

This bad behaviour is often re-inforced at the SCQD board and the daily meetings. Numbers are written in green if good and red if bad. This is ok for instant judgement of a particular result, it is good visual management. But once you see a red value and decide you are going to put some effort into investigating the situation, plot a graph of the last 90 days of results...and ask CHAOS to CONTROL?

If you have the right kind of data. Plot a **run chart**, plot a **histogram**, plot a **Cpk** chart and then look at the information and what they are telling you.

Now we have data being collected on the primary measure and maybe another level down what I would call secondary measures. But what are we going to do with it. At so many companies this is as far as they go. Just displaying and discussing pages of data or point estimates i.e. *'the reject rate this week is 4%'* can you tell me what this means? In some companies this would be an awful result, but what if this process has never seen such a low result? And the previous week the process delivered a reject rate of 12% now how does that feel? Data is meaningless unless it has context and the tools that deliver that context are the graphical tools:

- **Run Chart**
- **Pareto Diagram**
- **Histogram**
- **Cpk Diagram**

And if it helps get more knowledge from your data:

- **Box Plot**
- **Multi-vari Chart**

These charts are designed the turn data into information and in most companies we have too much of the former and not enough of the latter. These are also the only charts I use. They are simple straight forward and they work. Avoid the being seduced by the Chart wizard in Excel.

Now we have data we can describe the use of these tools.

The Run Chart

This is the chart that you will use to plot your primary measure. You will update it every day and look at it every day.

It is surely one of the most underrated of all the charts, its too simple isn't it to just plot the data points on a line graph in time order? I could see the same thing from looking at the numbers, couldn't I?

No you can't...

I recently visited a long-standing client to help on a project and move it forward. As part of the day we stood for about 2 hours at a Grinding machine, talking with the operator. He showed us what he did, measuring 100% of the product and adjusting the machine as necessary. We looked at every data point he measured, but we didn't write anything down or create a run chart on a piece of paper. After a few hours we wandered back into the office and I asked the two engineers a simple question. Was the Grinding machine set on the target? Was it set on the nominal?

We'd watched this machine for 2 hours, watched every part, every measurement and was it on nominal? We looked at one another and we didn't know! And best of all I'm pretty sure the operator didn't know either!

All he needed to do was plot a graph, no computers or technology. Just a piece of graph paper and a pencil. That was the basic advice I gave them, next step. Stand with him and plot a graph. The graph is below...

Just for you to know, the top edge of the graph is top spec, the bottom edge of the graph the bottom spec and each data point has 2 results. The max bore size and the min bore size observed by turning the part on the measuring mandrel. We have plotted a graph through the midpoint of each pair.

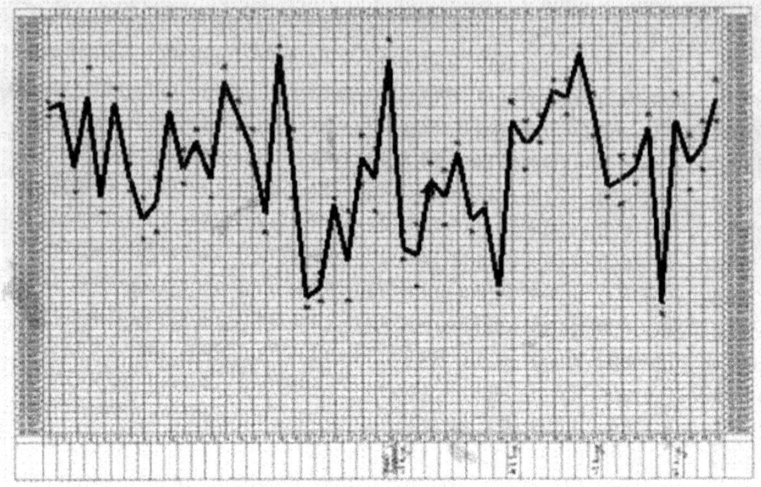

This process is nicely capable and sitting well inside the limits and yet it makes defects?? And the operator felt the need to make 4 adjustments during the production of these 50 pieces. To be fair to the operator he couldn't see this pattern as he worked, this is just a record of what he did normally, we created this graph, but he never used graphs or saw this pattern.

Should this process make defects?

If the operator plotted this graph, I would expect them to do a couple of things. Centre the result as they are running a little too high and over size in this case means scrap, not rework. Then once the process is centred take your hands off, drink tea and read the paper as the machine makes good parts all day! The process is perfectly capable of sitting inside the tolerances with no adjustment needed.

The only reason he doesn't do that currently is he doesn't plot a graph!!

And this costs his company over £10,000 a year in defects and the cost to fix it?

A pencil and piece of graph paper!

Always, always plot a run chart. If the operator measures something, plot the graph! Immediately, at the point of activity.

The Pareto Chart

Based on the 80/20 rule. I.E 80% of the wealth is owned by 20% of the people. In a project 80% of a reject rate is often caused by just 20% of the failure modes. This is the ideal use of this tool. A reject rate based on failure modes as opposed to missing a specification. When you are counting rejects and there are several different ways for a parts or service to fail, this is count, discrete or categorical data. The number of each reject type is represented on a bar chart in order, highest to lowest (see below). The idea being to allow you to focus your project on the most frequent failure mode. In the chart below diode insert, output, HV pin insert create 80% of the reject rate over the last month.

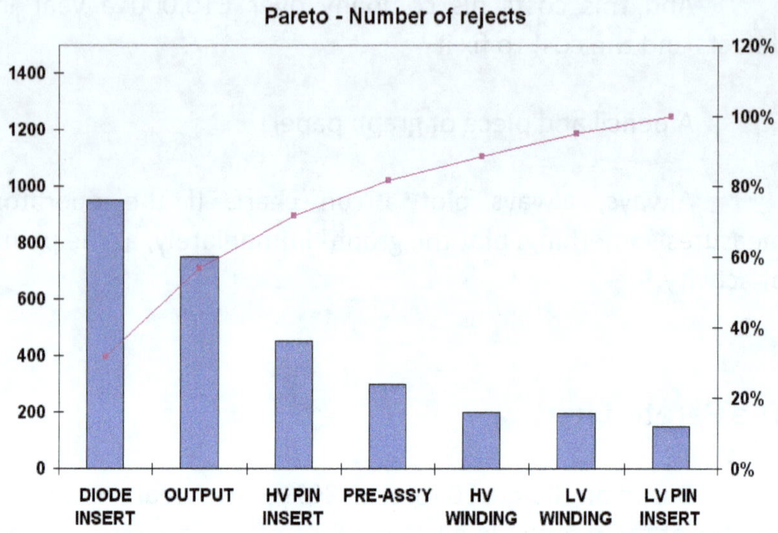

But again the key to this is not to teach you how to do a Pareto as it's a technique most people know about. An important point though is the timeframe over which data is collected and then charted. As I've just mentioned this data is count data and for count data to estimate what is happening accurately it needs a good sample size. In this case between 1000-3000 data points. Whereas the run chart should be plotted most often, usually daily. The opposite is true for the Pareto, the timeframe must be at least a month, not weekly and definitely not daily!

This is because count data is inherently in-efficient in delivering information and therefore needs high sample size (collect lots of data). If you try to use the chart daily you simple don't see the real signal telling you which is the biggest chronic issue, you just see noise as the chart changes daily without anything really being done to fix yesterday's problem.

This is a very common problem to think that we can fix a problem in a day or short time period. Refer to run charts and histograms on patterns that will tell you when it is possible to fix a problem quickly (which is not very often), most problems are medium term and this supports the need for a pareto to have at least 1000 data points to see real signals and to give you chance to do something about the issue before looking again at your biggest problem.

In order to create this type of chart you're going to have to create a simple data collection sheet. Possible a simple 5 bar gate as below.

Failure Mode	Daily Count
Diode Insert	𝍸 𝍸 ||||
Output	𝍸 |
HV Pin	𝍸 |||
Pre-Assy	|||
HV Winding	𝍸 |
LV Winding	𝍸 𝍸 |
LV Pin Insert	𝍸 ||

One last point about the pareto, not so much about the chart, but the way the data that is collected. A common mistake is to have too many categories of failure mode. I have been to some clients where they have over 140 different failure categories. If you draw a pareto in this situation nothing looks very important to fix, so you never attempt process improvement. A good practice to get into, is to count the number of times a process lets you down. Then plot that process count on a Pareto rather than the initial reject modes.

This will make you concentrate on the Process rather than the product. To improve a process is a much faster way to improve than to fix each individual mistake, which is a very slow way to improve.

Another mistake is to convert the failure into an assumed cause. An example might be a paint defect, its small speck of dust in the paint. Instead of recording it this way the operator looks at and records the cause to be dirty hangers. That's a guess!!

It would be better for him/her to simply record dust spec as a failure mode and if you get enough of those, you can start a project and do some process analysis/improvement work.

The Histogram

A histogram is a specialised type of bar chart, individual data points are grouped together in classes, so that you can get an idea of how frequently data in each class occurs. The strength of the histogram is that it provides an easy to read picture of the location and the variation in a data set. So look at the data below. 50 data points from the performance of a product. What do you see? Numbers right? Now let's create a histogram of the data.

49.9	52.3	50.7	51.1	51
50.6	51.7	50.9	51.8	51.8
50.8	51.5	50.7	51.4	49.4
50.3	50.1	51.9	50.4	50.2
52.8	49.4	51.5	50.1	51
48	50.3	50.6	51.1	49.8
51	51.9	51.7	51.8	50
51.3	51.5	52.5	51.2	50.7

50.8	51.4	52.4	51.4	51
50.1	50.1	51.3	50.5	52.3

Now we just count the number of data points that fall into each group and the distribution below is formed. Visually its easy to see the location, or mean, I prefer to call this the signal, (the mean is also an estimate of the centre) you can also see the spread, often measured by the standard deviation, which I will call the noise and often obscures the signal. This picture also allows us to see the distribution which is a normal shape and we can use this to predict reject rates, without having to collect huge amount of data.

For instance, if we drew the specs on the histogram for this process, we would get a sense of the reject rate immediately and unless this process performance is changed that reject rate will exist day after day.

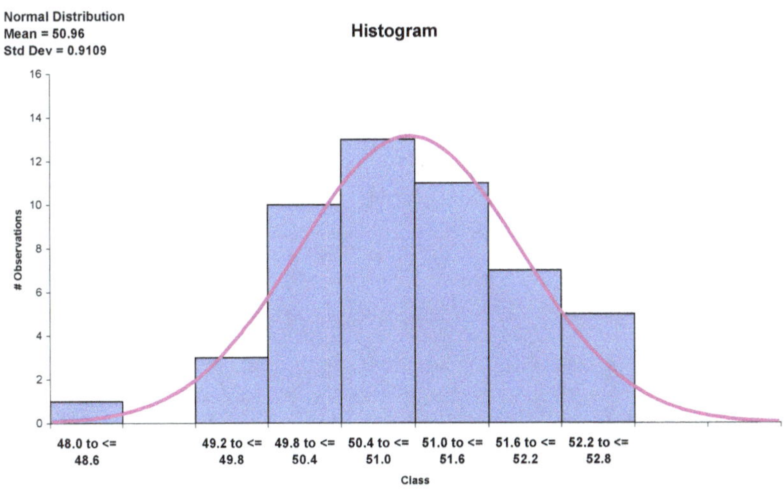

The histogram is a great example of why you always look at data through the lens of a chart rather than looking directly at the numbers. It shows a pattern that you cannot see in the numbers alone.

The picture created by the histogram is so important to being able to understand what to do next. Look at the 4 scenarios' below. Each one is different; each one has to be tackled in a different way. Each one will take a different amount of time and resource to fix.

Example 1

Below is a histogram of a process that we've been having a problem with. I've added spec lines to it to help add extra information and insight.

Looking at Histogram 1 below we can now see that the process has too much variability for it to produce within the specification required of it. In other words, this process is not capable.

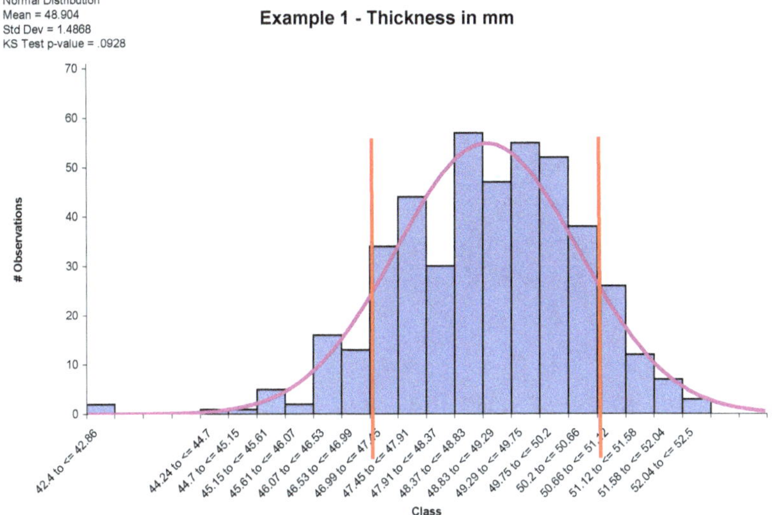

To fix this problem we will need a technique described in Chapter 2. Process Flow, Cause and Effect analysis. Find all the variables and their current state of control. We will then control all the input variables that are not currently controlled to make the process as capable as possible.

This is going to need a team of people and solving the problem will take months to achieve and only management can make this happen. To be honest this process should never have been allowed to run. This problem should have been spotted at New product introduction. How often do you set up processes like this in your company and are then surprised when you get continued defects?

One last important point. It cannot be fixed by the local technician on the machine adjusting the process. In fact, if the person on the machine tries to fix this problem by making process adjustments to move the centre of the process back and forth the only thing they will achieve is to make the defect rate worse! Show the technician this picture and tell them to sit on their hands, drink tea and read the paper, they cannot help, other than to grade good parts from bad.

They say that 90% of process problems are like this and cannot be improved by the operator. But 90% of the time the operator is trying to fix the problem and making it worse!!!.....

Example 2

This process also has a problem and is regularly producing defects. However, we can see that in general terms the process is capable, happily producing results which sit between the specifications required of it. However, now and again it produces a complete outlier, something totally different to the natural capability.

This is a sign that there is a root cause, a single thing that is going wrong. This will need some kind of in-depth analysis of the process to find......a Standard operating procedure is being ignored, material that has a defect etc a single fault.

This can often be fixed relatively quickly and once fixed the defects will go away. It might be good idea to get opinions from a cross section of people, but it is not always necessary to use a team for this type of problem. In this case it might take a matter of weeks to fix...

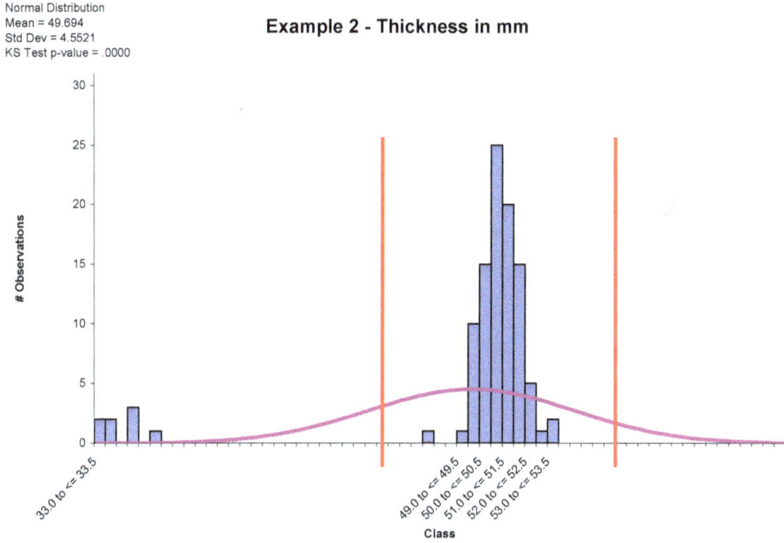

Example 3

This is the simplest of all problems to fix. The process is capable of falling inside the specifications, but it simply needs to be centred. This problem can often be fixed by one person, often the person running the machine by adjusting a setting on the machine. In which case, this problem could be solved in minutes.

If moving the mean is slightly more technical, getting a tool modified for example then the solution might take a few weeks, but certainly no longer than that. This type of problem never needs a team to fix it.

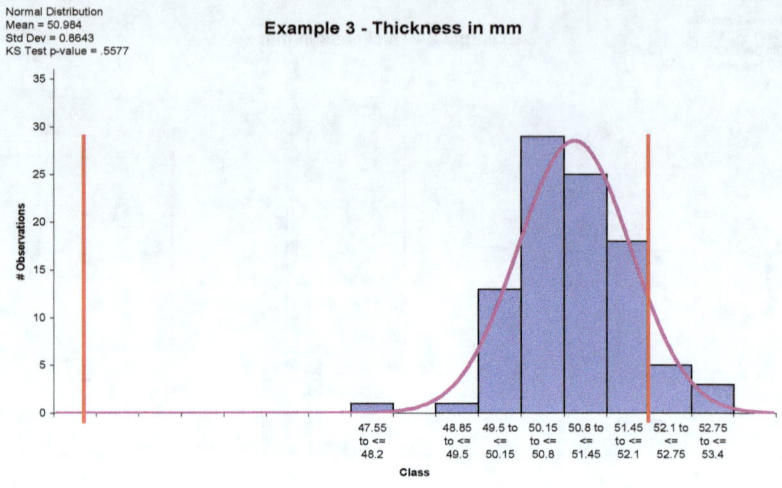

Example 4

A great example of chaos, no obvious pattern and data is all over the place. This will need a team and correcting this will take months, only management can sanction this…

Sometimes the statisticians encourage us to make a mathematical transformation to make this look normal. Ignore this nonsense, this is a pattern that your process is producing don't corrupt it or transform it. The histogram looks a mess because your process control is a mess!

If you want it transformed, transform the process not the data!

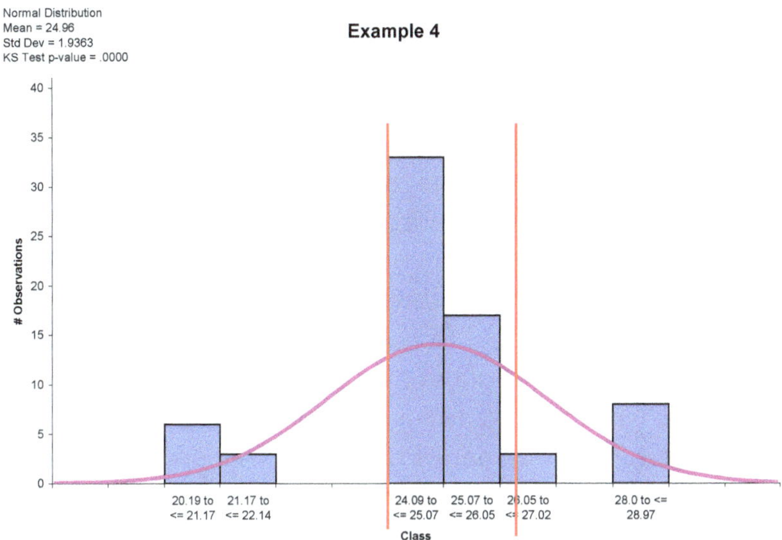

Summary Statistics and Statistical thinking

There is a great quote from H.G Wells "Statistical thinking will one day be as necessary for efficient citizenship as the ability to read and write". It's a paraphrase apparently and what he really said was to understand averages and Maxima and Minima.

Averages are measuring the middle of a set of data or process result, Maxima and minima are measuring the spread of the results. Another way of describing this is that the average or middle can be considered the signal of a process, maxima and minima are the noise of the process and it's only by considering both the **signal** and the **noise** that you can be said to be thinking statistically. So let's consider each and look at some charts that got with them..

Measures of Location – the signal

To go with the picture of the distribution we get from a histogram we also have some statistical measures, the mean, is the measure of the middle of the data and the standard deviation a measure of variation or spread.

The measure of location that we all understand in common language is said to be the average; this is essentially an estimate of the middle of the data set. Although in the UK we use the word average to suggest that we add up all the data and divide by the number of data points, this is in fact more correctly known as the mean, one of 3 basic types of average. 3 estimates of the middle.

Mean – Add all the data and divide by the number of data points.

Median – The middle number in a list of numbers when they are put in order.

These are different estimates of the middle. Once again, it's an estimate and when using them you must ask that sensible question. *'is it a good estimate'?*

One of the big issues with the Mean is its sensitivity to an outlier, which can dramatically pull the estimate well off centre and create a poor estimate.

It's perfectly sensible to remove these from the estimate calculation. That is not to ignore them, especially if they are reject results. They are after all extreme events they need to be investigated and the causes eliminated. But they shouldn't be included in the calculation of the estimate of the middle. In Example 2 from page 85 we have an example of this. The centre of the main set of data can be clearly located just by eye. The extreme values out to the left however, would pull the calculated value into a very strange place indeed. Clearly these events need to be investigated and eliminated, but they shouldn't be used to calculate the mean.

You can see this on the histogram below the mean has been calculated from all the data including the outliers and the mean is 49...

The natural centre of that normal distribution sitting dead centre of the specifications is nearer to 51...

In this case the estimate of the middle is not good if the outliers are included....

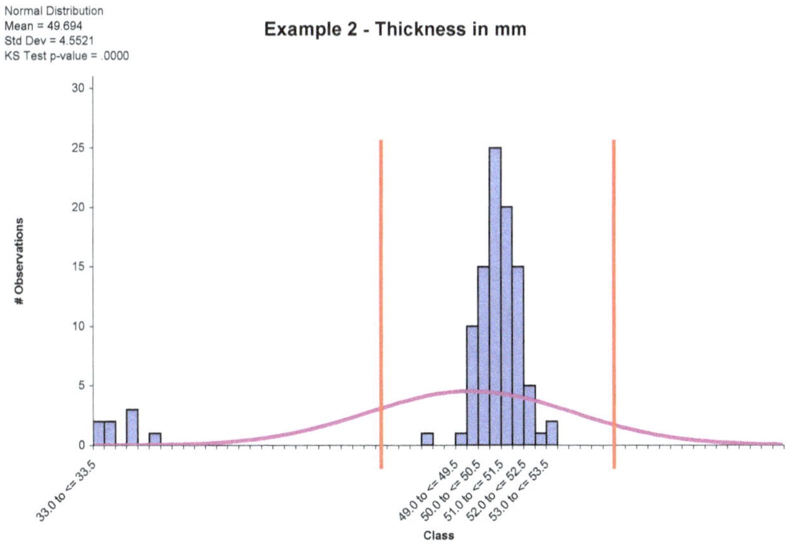

This is particularly important if the values are collected automatically and then fed back into the process for adjustments to be made.

In that case it maybe better to use the Median as an estimate of the middle as this is very robust to outlier values. It will also be in a similar place as the mean (outliers removed) if the distribution is normal with an even shape. So it automatically factors out the effect of outliers without having to 'correct' the data. The Median is simply the middle number once all the data points have been put into a list highest to lowest.

The Multi–vari chart

This is often, but not always, another time phased chart. It's also a chart that is only used for process analysis. It's not the kind of thing that would be produced at the point of activity or as a weekly/daily performance graph.

The multi-vari chart begins to segregate the data by different time periods or different variables, hence the name, multi-vari. This chart is often used when a dimensional problem is being looked at in the project. The chart relies on variable data being collected in a particular way. So the ideal way to collect the data is to take a small sample say 3 pieces which have been consecutively produced, these are taken every hour. Each time measure the dimension of interest in 3 different places on each piece. So if we were interested in the diameter of a shaft we would measure it not once but in 3 specific places, each end and in the middle perhaps....

With this data we can then construct a multi-vari chart that will quantify were the biggest sources of variability are coming from.

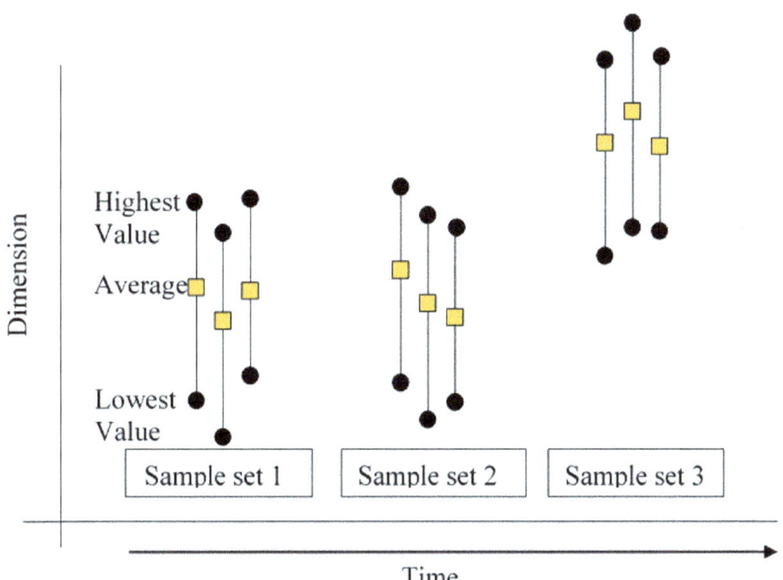

Each piece has been measured 3 times and each line represents this data on each piece. The top point is the highest of the 3 readings. The bottom, the lowest and the mid-point, the average. The chart will show variability in 3 times frames:

1. On piece variability - shown by the length of each individual line.
2. Between piece variability – shown by how much the average point moves up and down within each sub-group – the yellow squares
3. Group to group variability – Shown by how much the sub-groups vari between each other

In the case of the chart above the most variability is coming from on each piece and the between group variability as shown by the movement of the 3rd sample. The piece-by-piece results, the centre squares move very little. If you were analysing variables for this process you could now discuss which variables would have an effect in this time frame. Quickly eliminating other variables, which are less significant and giving the group focus.

This is a chart not covered by the SPC XL software although the box plot is very similar. But the box plot can only be used when you have 5 data points or more to create the box, which only happens in specialist situations. If you deliberately collected 3 to 5 data points for the multi-vari analysis I recommend you use the stock graph – High, low, close graph in the excel wizard, creating 3 columns of data the maximum result minimum result and the average result.

Multi-Vari - https://youtu.be/PWA5mn0nxwY

Apart from time phasing the data, another good use of this chart is to separate out elements of known variation. The example below demonstrates the use of the chart in a moulding situation and separates out the variation caused by each cavity.

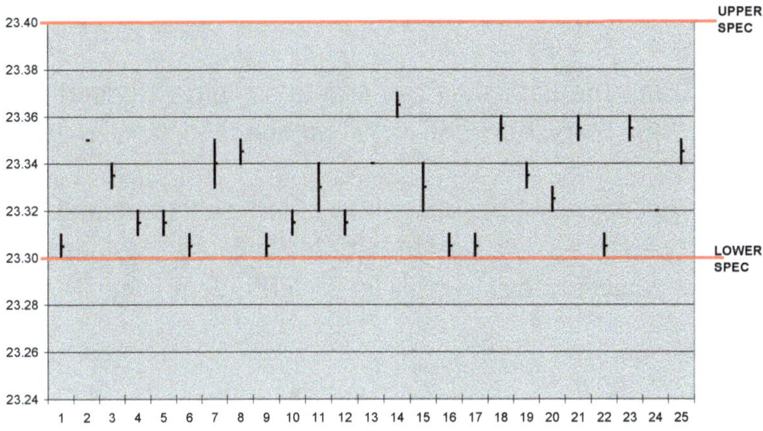

A quick look at the chart (Stock Graph in Excel) which has the 25 cavities across the bottom of the chart each line has been created from 3 data points per cavity we can see cavities 1, 6, 9, 16, 17 and 22 sitting on bottom tolerance these need immediate action. Cavities 4,5,10,12 are also sitting low in the spec so modification here would be helpful as well. This is a typical use of the multi-vari chart. If you dig a little deeper on Wikipedia, you will find more advanced ways to show even more information but for this text this is as far as we will go.

How to create one - https://youtu.be/JkytzK2HejM

Finally let's consider a final example for the Multi-Vari. A waste collection business wants to understand the variability of the waste arriving at a refuse station. Is the variability within container, between container or between days. Understanding how the variability arrives would help with decisions about how to run the site and how to store or handle the waste as it arrives.

They take 3 samples per lorry. They sample 5 lorries in a day and they do this everyday.

Using the data they can create a multi-vari chart that shows within Lorry, between lorry and between day variability.

That sampling decision allows them to understand how variability arrives at the station but only because they designed the data collection to show this up on a Multi-vari chart.

The Box plot

As we mentioned earlier, there are 2 analytical diagrams that can represent data as within part and between part or within day and between day data. We've already looked at the Multi-vari, I would use that when manually collecting data as it keeps the data collection to a minimum.

However, there are times when we get so much data from a system that we're buried in it and have no way to easily summarise and understand it. This when the box plot in SPC XL is a great tool.

A good example of this data overload is when a steel works monitors emissions from a chimney every 5 seconds to demonstrate compliance to an operating licence. They collect 17,280 data points everyday. In this situation a run chart makes no sense. And there is too much data for the multi-vari chart.

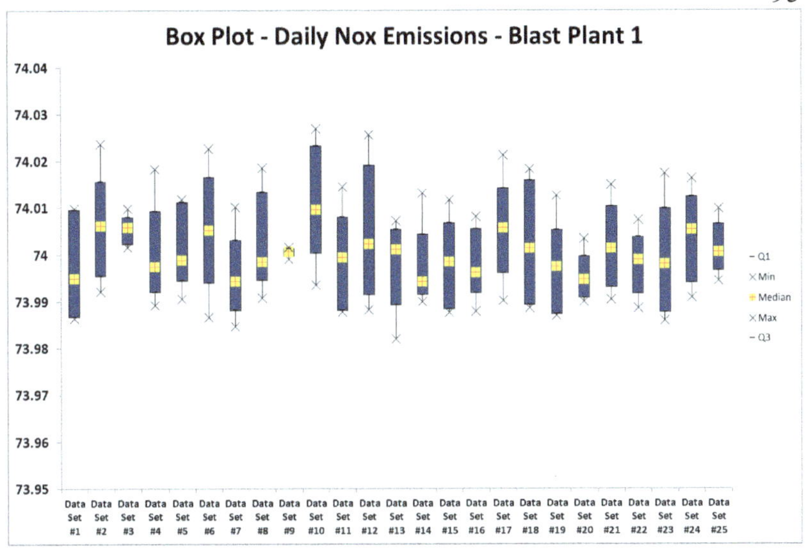

In the box plot example above each box represents those 17,000 data points. The cross at the top shows the highest reading that day, the bottom the minimum that day and the dark block is the middle 50% of your data. The highlighted cross in the small box in the middle of the dark block shows you the median value of your data. This is the estimate of the middle of all the data for that day. This one graph is summarising over 400,000 data points and giving you crucial information about within day and between day variability. No other graph in the set we are looking at can do this so well and so easily.

Measures of Spread or dispersion – The Noise

Taking H.G. Wells quote about the maxima and the minima the obvious measure of noise in a process would be the range. It's a number which is very easy to calculate and easily understood by even the most nonmathematical person. It is certainly a value I will use as a quick way of understanding variation. The range is essentially the total variability in a process data set.

However, given time to make a calculation there is a much better measure of variability. That is the preferred measure. It is in fact the average variability of a process, and it is known as Standard Deviation. It's a number we learned to calculate at school to pass an exam at 16 and then quickly forgot all about it after that. I know I did.

The formula for it is shown below..

$$s = \sqrt{s^2} = \sqrt{\frac{\sum_{i=1}^{n}(x_i - \bar{x})^2}{n-1}}$$

Looks awful doesn't it? No wonder no one wants to use it. However, it is the most brilliant number to know! Because unlike the range, once you have the Standard Deviation for a process it is possible to predict the defect rate without switching the machine on. The real power of statistics....prediction!!

If you look at the normal distribution below, a pattern that is very common in process results. There is a link between the amount of data we expect to fall, one standard deviation (σ) from the Mean, two Standard deviations (2σ) and three Standard Deviation (3σ) from the mean. In fact, wherever I draw a line on that diagram I could predict (estimate) how much data will fall to the left or right of that point. Of course, if that line happens to be your specification, then you are estimating the defect rate for your process. This can be done with only 30 to 50 data points and will predict the defect rate even if the sample of 30 contains no defects at all. Think about that for a second, with only 30 data points I can predict what the next 10,000 pieces will look like!

You can't do that with the range, that's why we use Standard deviation we can predict tomorrow!

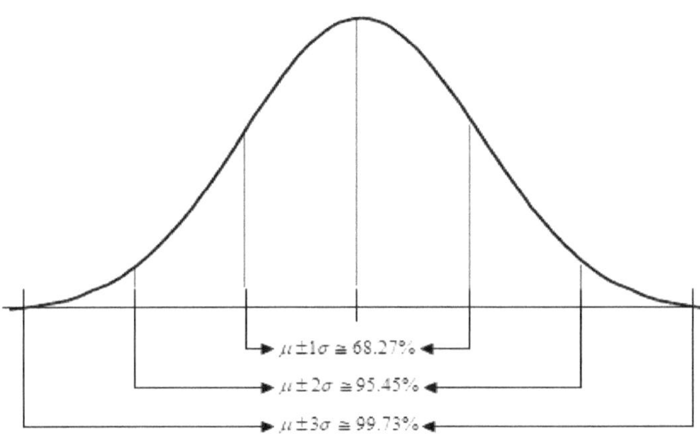

The Cpk Diagram

Having explained that to know the mean and the standard deviation for a process we can calculate and predict the defect rate for a process, that is exactly what we are going to do with the Cpk Diagram.

Cpk is essentially a process capability measure. Process capability just asks the question how well my process results fit in to the specifications.

It compares the voice of the process (the Data) to the voice of the customer (Specs).

I think it would be good to watch the video and then read the text and then watch the video again later.

CpK Video - https://www.youtube.com/watch?v=QH984PnwRDE

If you look at the diagram below. How well does the normal distribution fit inside the two lines? The specifications.

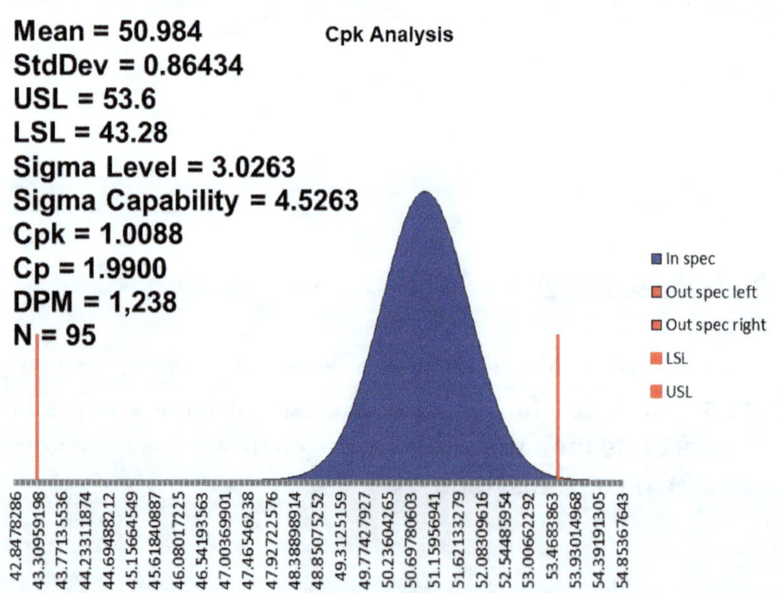

There are 2 capability measures that we use to describe what we can see in this picture are.

Cp – the process potential.
Cpk - the actual process capability.

The Cp the process potential just asks a very simple question.

How wide are my process result going to be? How much variability does it have? And how many times can I fit that width inside my specification lines. It's basically a width to width comparison.

If you look at the picture above in a very simplistic way. How many times can you fit the normal shape inside the lines?

In this case just by eye, you could fit 2 of those distributions side by side so Cp is 2 approximately. The calculated value for Cp is 1.99….

The way the calculation is done is pretty simple. Any normal distribution is said to be 6 standard deviations wide. The distribution then is 6 x the standard deviation and this is divided into the width of the Specification.

$$Cp = \frac{USL - LSL}{6X\sigma}$$

The important thing to realise about the Cp is that it is only the process potential and a width to width comparison. It is not concerned with the location of the distribution at all.

The distribution below is completely outside the specification making pretty much 100% defects and yet the Cp is the same 1.99.

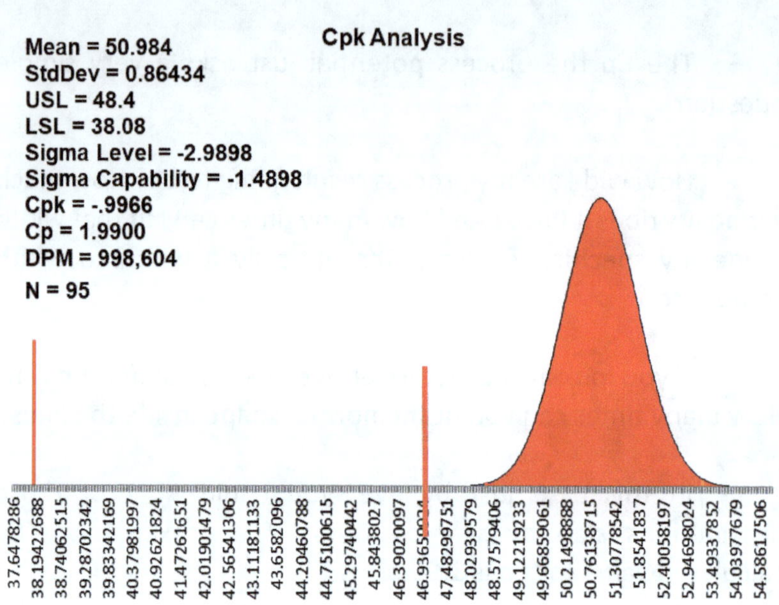

That is because it will still fit twice inside the specification limits as a simple ratio.

This value of 1.99 is the potential for the Cpk value, if you can get the process perfectly centred. If this is the potential for the capability, what is the actual capability?

Well that does take the location of the process results into account. To do that, its works out which specification the process mean is closest to. In the example below, the process is well off centre and much nearer to the right hand or upper specification. We are only really looking at half the specification. Therefore, the calculation for the Cpk compares that one-sided specification to half of the process or 3 times the standard deviation.

In the diagram below you can see half of the normal distribution (the process results) fits into that space between the mean and the right-hand spec approximately once.

The actual Cpk figure calculated is 1.009.

The potential, the Cp is 1.99 if you could centre the process this is the best your Cpk would be.

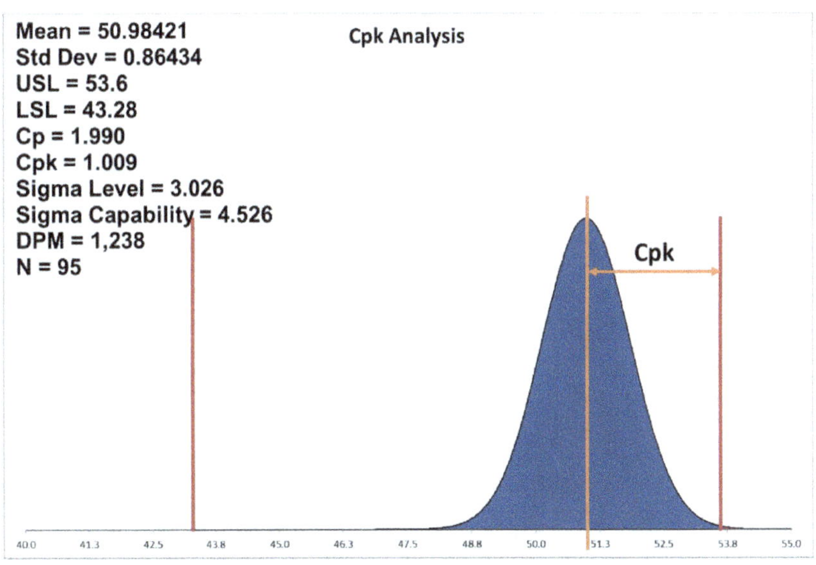

The example below, the process is more centred, and the Capability statistics are:

Cp = 1.3498
Cpk = 1.1630

These values are closer as the process is nearer the centre.

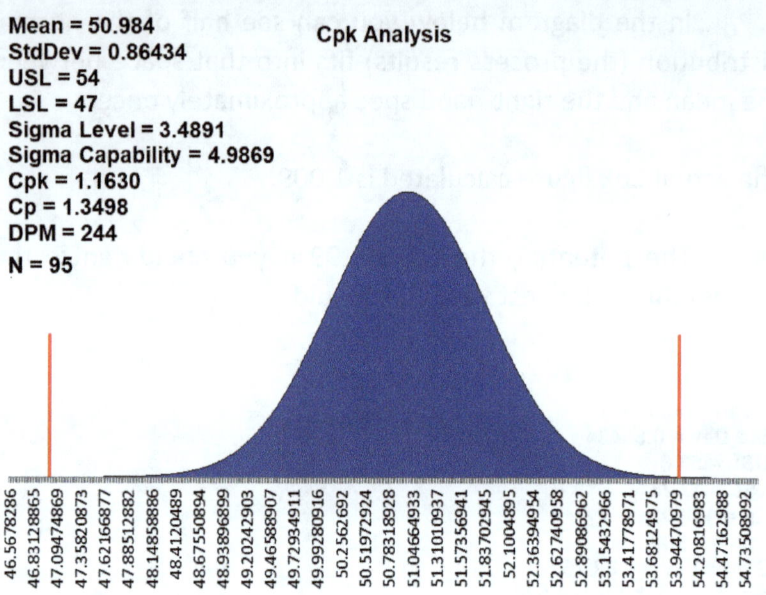

Now despite all these wonderful capability statistics, I have to be honest, the way I work with these diagrams is to look at the DPM, the defects per million and use my common sense.

The example below. Defect rate 44,052 DPM the problem is the spread of the process, the noise. Measured via the standard deviation. This will need a team to sort out and will take me 3 months to achieve using process flow/cause and effect. Do I need to look at the Cpk value to figure this out? Not really.

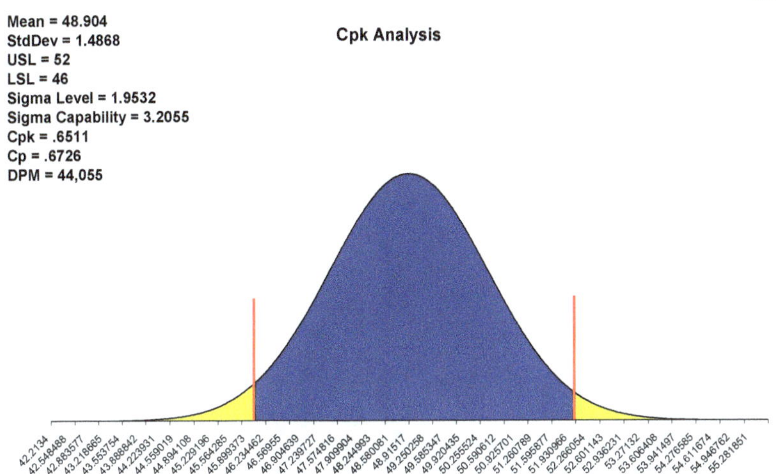

In the days when you couldn't draw these pictures very easily, the Cp and Cpk values were essential.

Example 2 below is a process off centre. The defect rate 119,954 DPM this can be quickly reduced just by centring the process. However, Would I be happy with that? Well now the Cpk and Cp are useful.

Cpk = .3917
Cp = 1.1569

If we centre the results, the process will fit inside the specs according to the Cp 1.15, however this does not give the process much room to get stressed, pushed off centre. Whilst you can quickly reduce that defect rate, by changing the signal, the mean. You might also want to work on the noise over the middle term and reduce the variability.

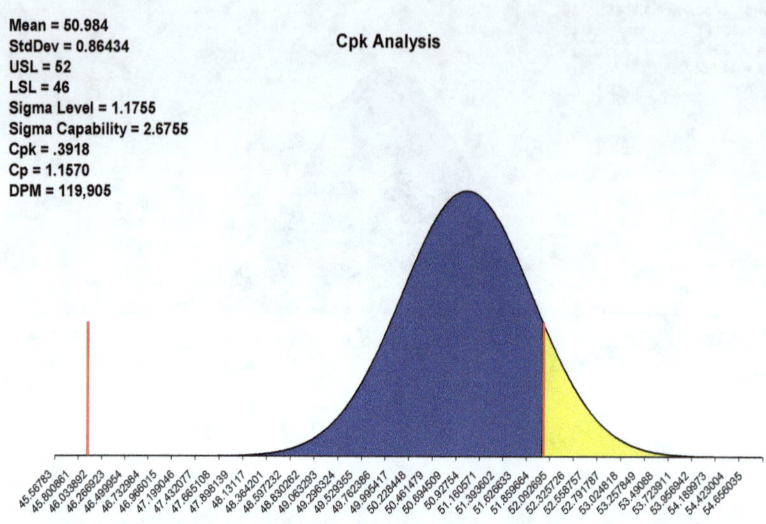

This again is a good example of thinking statistically; do I need to move the Mean or control the Standard Deviation. This is how we understand the process physics. Only by thinking about both of these statistics do we know how to react and what to do next and how long it will take. Moving the Mean can often be done in 3 minutes, 3 Hours or 3 days and can be done by one person. It is sometimes called a special cause variability. Controlling the Standard Deviation will take 3 months and needs a team and a project. This is common cause variability.

Measurement System Analysis (M.S.A.)

Although we are still in the Measurement Phase of your project, we have so far concentrated on how to take data that you collect and paint pictures and summarise it to learn so much more about the state of your process physics in order that you can plan your problem solving approach.

This section will bring into question the data that you have used to created that summary information. I think it could be argued that this section should have been first in this part of the text. But I wanted you to value the diagrams and statistics before we get a little bit more technical and start thinking about how to make sure your measurements are correct. So yes before you measure, please do an M.S.A on your measurement systems.

So even though we've made the analysis of data seem very straight forward, let us start by being a bit controversial, every measurement system you have is wrong! Your measurement is a guess or to be more scientific, an estimate. Now if you accept that every measurement is an estimate, what would be the next sensible question you would ask?

If my measurement is wrong, how wrong is it? If it is an estimate, how good an estimate?

If it is wrong by a small amount, then you are ok it's a system you can use and you can trust the information coming out of those diagrams. If the potential error is large, then you have a problem. In fact, it is a problem you must address before even thinking about whether you have a manufacturing or process problem.

In most organisations they accept the vulnerability of the measuring process because they have already put controls in place to ensure error is corrected or minimised because they calibrate on a regular basis.

Error in a measurement system can come in different forms.

1. **Accuracy** or should I say inaccuracy
2. **Bias** – Under or over measuring consistently
3. **Precision** – Not being able to reproduce the same measurement for the same part.

When we are looking at variable data, data measured on a scale, **accuracy** and **bias** are the domain of the calibration process and make sure the equipment is set and ready to use.

Calibration often entails the measurement of perfectly formed simply shaped reference components in perfect conditions. Unfortunately, not many businesses produce perfectly formed simply shaped parts. It is the use of a measuring system in real situations where the problem of **precision** lies and precision should be assessed for every different job a measurement system is used for. This is what MSA will do for you.

A single measurement system would have one calibration, carried out once or twice per year. but if that System is used to measure 10 different components then it would have 10 MSA's conducted for each part/dimension it measures.

MSA therefore is a practical tool, to be used often by the process engineer, to see if the measurement system works in practice. The calibration procedure checks that a measurement system works in theory and is the domain of the Standards Department or often an external calibration service.

Everyone in manufacturing has probably come across the symptoms of measurement system error. You measure a batch of parts once and they just fail, the result is just outside the tolerance. So, because the results were close to the tolerance, what do you do?

You measure them again, right?

This time they just pass. Isn't it funny how we are always ready to accept the second measurement as the one which is correct, and that the first measurement was in error? Or you send parts to your customer. They measure them and decide they are outside spec. You get them back and re-measure them and find they are OK, and you can never agree who is right.

Both events are sure signs of measurement error.

Now I am sure you will jump to the defence of your measurement systems by saying *'they can't be wrong they are calibrated!'* and that is a good point.

But calibration only controls the equipment or machine variables in your measurement system and when we use the phrase measurement system, we are not just talking about the measuring equipment. We are also including:

the people
the environment
the parts that are being measured
the methods being used.

For example, if a £20,000 highly accurate ruby tipped, computer-controlled co-ordinate measuring machine (CMM) tried to measure the diameter of the pin that is oval, what is the correct answer? Remember at this level of accuracy you would not see the ovality in the part by eye. But even assuming you knew that it was oval, what would you do to get the 'correct' result?

Is there a correct result?

You could measure the same pin 10 times and get 10 different answers. This is lack of precision. This error created by the shape of the part being measured, is part of the measurement system and part of the error you see when trying to evaluate if the Process is set correctly and OK to run. Or part of what you see when trying to monitor the process during a run.

You could agree to measure the component twice at 90-degree intervals and average the results, or measure in 3 places and average the results. This is variation in the method and is also part of the measurement error. Each method will produce a different amount of measurement error and effect the accuracy of your estimate. And without a Standard Operating Procedure, different operators will undoubtedly use different methods.

If we accept that the measurements you take are only estimates, as the system you use cannot be any more accurate than that. We really need a method that will quantify how good an estimate is it? And to quantify how much this error is adding to the variability in your process results and how much that inflates your defect rate, that is one of the consequences of measurement error inflating your defect rates.

Measurement System Analysis will provide the information to tell this.

There are 2 events where you could use an MSA. One is to use it in your problem-solving procedure. In a worst-case scenario, the problem you are investigating might be exclusively coming from the measurement system and we have a case study later in the course to show you exactly that. Or you could use the MSA as part of your process development when conducting New Product Introduction. In the official Production Part Approval Process (PPAP)

And there are essentially two types of measurement systems analysis that you can undertake. The first is when you are measuring variable data, something on a scale like millimetres, volts, grams, air flow. The second is pass/fail in nature such as a cosmetic inspection technique or when you are categorising defects. Is the fault a dent or a scratched for example...?

Because we have the use of some suitable software, some of maths in these 2 methods will be skipped over as it adds no value to demonstrate the hand calculations. The diagrams and graphs that you will see through this text has been created by **SPC XL**, a 10 Day trial version of this software is available from sigmazone.com

The technique that assesses variable systems is often called Gauge R&R and that is the approach we shall discuss first after watching the video that goes with this section.

MSA Description Video

https://www.youtube.com/watch?app=desktop&v=23ALp0CFImE

Chapter 6

Measurement System Analysis for Variable Data

MSA for variable data is known as Gage R&R. The R&R in the title stands for **repeatability** and **reproducibility**. These are the 2 basic types of error you get in a measurement system.

Repeatability is seen when a system cannot repeat its own measurement on the same part (**within** system error).

Reproducibility is seen when 2 systems disagree and cannot produce the same result for the same part (**between** system error).

The study will be conducted on a 'normal' sample of parts, usually 10 parts, randomly selected from your process. That fact that you select these parts randomly is very important. If you pre-select parts you will ruin the MSA. The other important point is that the parts are of one feature on one part only. Again, mixing sizes or products will ruin the results

Before you conduct this type of study you need to be clear what questions we are trying to answer by making this analysis. There are 3 basic questions.

1. Is my measurement system any good?
2. If the answer is no to question 1, where is the problem?
3. What am I going to do about it?

When looking at variable data, measurement error essentially adds variability to the process results. If we look at data from your process as a run chart initially then we get the pattern below.

Measurement error is making the peaks and troughs more extreme. Meaning that you could be getting closer to or even crossing the tolerances making it look like you are producing more defects than you really are.

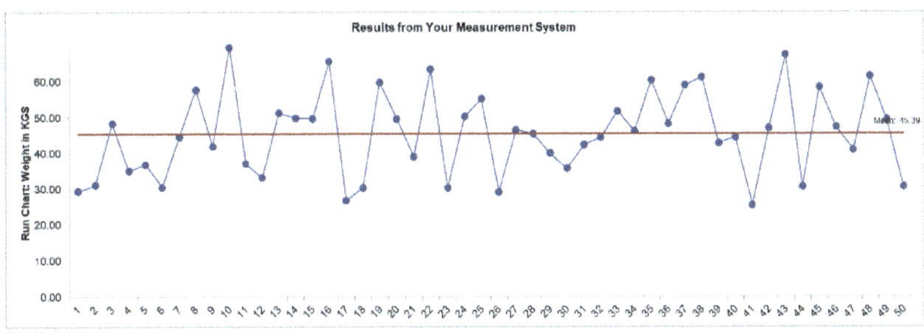

If you turn that around to look at the distribution that is being created it would look like the diagram below. We would like to see the real results coming out of the manufacturing process (identified as the narrower product distribution). Unfortunately, you can never see this manufactured result. Because, before you see this pattern the measurement system adds variability to it and makes it look 'fatter', more variable than it really is.

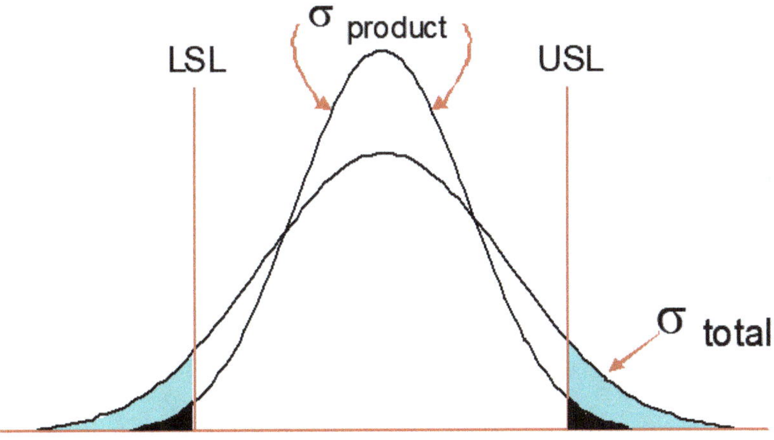

This inflates your defect rate slightly (hopefully only slightly). What an MSA will do is tell you how much the manufacturing results

are being inflated. The difference between the results you observe (sigma total) and the real underlying distribution (sigma product)

The R&R study is conducted to get the maximum amount of information from the minimum amount of data. A statistical sample size essentially.

The strategy is to include all the operators, equipment, and methods etc that are normal to the measurement system.

A Random selection of parts are taken to represent the inherent process variation. This is very important, the selection of parts used in the assessment must be random. If you throw out defects etc or pre-select the parts in any way, the test will be affected. Usually, 10 are selected. These 10 will be measured multiple times and if you are only assessing the minimum of 2 Systems will ultimately will produce a minimum 40 data points,

The Parts are labelled in such a way to remove operator bias. This means blind marking. Parts are going to be offered to each measurement system or person twice. Of course, if you are asked to measure the same part twice, knowing that it is the same part what answer will you get? Hopefully, the same measurement (try it, it never is the same).

If an operator knows that they have seen this part before and can remember the previous results they will make sure they repeat that number. This is a biased result...

A gauge R & R study is going answer the first of our 3 questions by generating two important statistics.

Precision to tolerance Ratio - 0.10 or 10% is acceptable

and

Precision to total ratio - 0.10 or 10% is acceptable

Precision to tolerance tells you how much of your specification is being used up by your measurement system.

0.16 = 16%
0.45 = 45% .... of your specification is being used by measurement error.

Precision to total ratio tells you how much of the total variability you see is coming from the measurement system. How much the original results are being inflated by measurement error. And the results can be looked at in a similar way

0.21 = 21%
0.38 = 38% .... of the variability you are seeing is coming from your measurement system.

These statistics are shown on the data page of the MSA from SPC XL

MSA ANOVA Method Results

Source	Variance	Standard Deviation	% Contribution
Total Measurement (Gage)	5.577E-07	0.00074679	49.33%
Repeatability	5.5255E-07	0.000743338	48.88%
Reproducibility	5.1446E-09	7.17256E-05	0.46%
Operator	5.1446E-09	7.17256E-05	0.46%
Oper * Part Interaction			
Product (Part-to-Part)	5.7282E-07	0.000756849	50.67%
Total	1.1305E-06	0.001063257	100.00%

USL	0.009
LSL	0
Precision to Tolerance Ratio	0.49786012
Precision to Total Ratio	0.70236076
Resolution	1.4

In the example shown the....

Precision to tolerance is 0.49 (49% of the tolerance is used by measurement error)

Precision to total is 0.70 (70% of all the variability I see in my data is measurement error).

So, question 1

Is my measurement system any good? The answer NO!

Now we need to answer question 2

Where is my problem?

To answer this question, we are going to be looking at some other statistics and the graphs produced by SPC XL....

Firstly, look at the Standard deviation **reproducibility** and **repeatability** which is the biggest? You can see the figures highlighted in the Standard Deviation data column above. That is all we need to consider...

In the case above the **reproducibility** Standard Deviation is 10 times larger. Therefore, the error is due to the differences **between** systems not within each system.

Now we can look at the graphs to help us understand this better. Below is plot of the average results that each operator measured for the 20 parts.

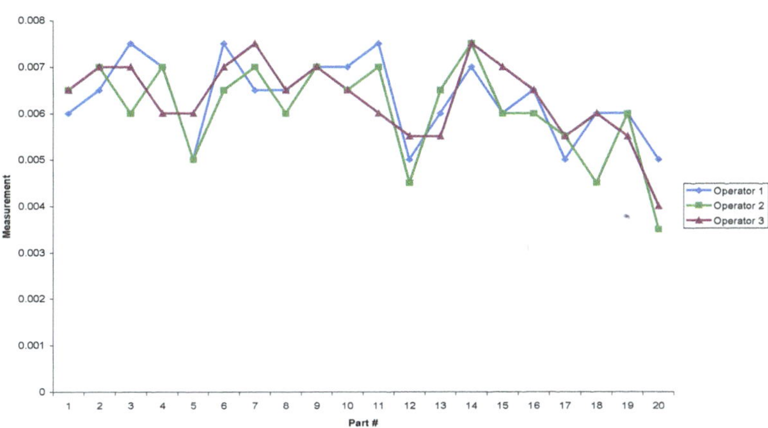

You can see from the graph above two patterns that would of concern to you. Firstly, the differing results from each of the 3 operators for the same part and secondly that there is no pattern that each is following. After all, it would make sense that if part 2 is bigger than part 1 you would expect all the graphs to show this pattern. Part 3 for example is particularly bad. One operator thinks it is bigger than part 2 whilst one operator thinks it's smaller. This is clearly a cause for concern and usually indicates no consistent process for measuring system to system. The measurement process is a complete guess!

The range chart below on the other hand, which is plotting the difference between the 1st and 2nd measurement for each operator shows little difference between the performance of operators. A good result on this graph would be all the results at 0, which would indicate there is no difference between the 1st and 2nd measurements of the same part, each operator can repeat their individual results.

In this case they are all getting a similar error as they try to repeat a measurement for a single part. The other important point is to look at the scale on the left-hand side, in this case the scale of the biggest error is 0.002mm, so even though we have a small amount of within system error **(repeatability)** in the calculated table it is relatively small and these graphs just reinforce what the statistics are telling you, the biggest problem is **reproducibility**, each system cannot agree with the other systems **between** system error.

These graphs also enable you to answer the last question.

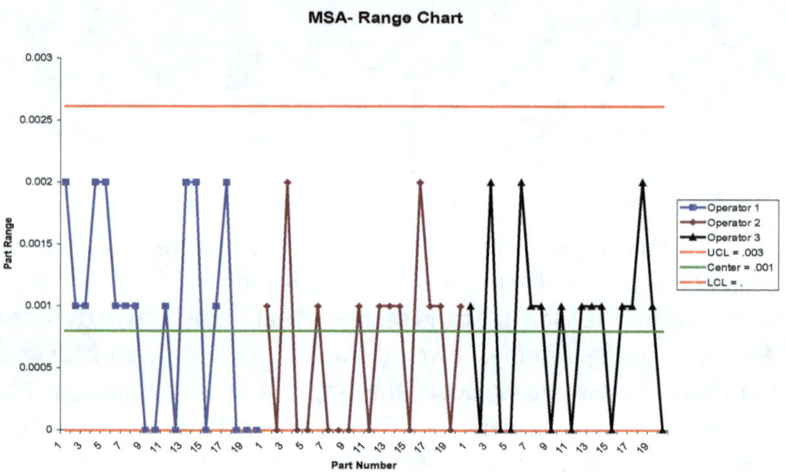

What am I going to do about my problem?

In this case because the operator by part graph was so random, no systems were agreeing on a measurement at all, and there is no obvious difference between operators in the range chart, a good specific SOP for the measurement process is needed. Get it written then retry the measurement system.

If Operator (or system) 2 or the 3 where agreeing, then perhaps they are following an agreed SOP and the 3rd operator needs training. Interviewing or asking the operators would always be a good thing to do at this point and involve them in writing the SOP.

In SPC XL you will get more graphs to look at than the ones I have just referred to. As a reminder to answer the 3 questions posed in an MSA I would look at the statistics table, look at the first graph on the Excel worksheets (Operator by Part). And look at the last graph (the range Chart).

The only other useful diagram I refer to looks like the one below and it calculates the misclassification rate if you continue to use this measurement system. In other words, when a part is close to the tolerance, will the measurement error move it to the other side of the tolerance and therefore misclassify good parts as bad and vice versa. It does not help to decide if the system is any good, but it does tell you how important it is to get on and solve the problem or face unhappy customers potentially.

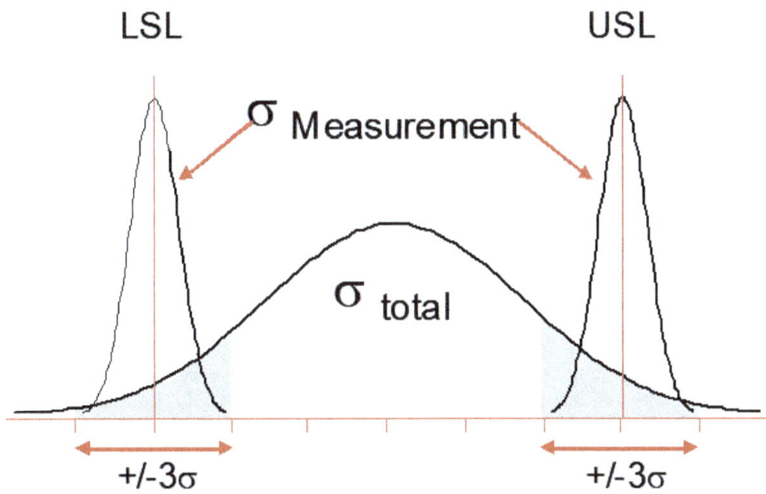

Chapter 7

Measurement System Analysis for Attribute Data

So far, we have looked at the error that occurs in measured or variable data. But of course, we also have errors in pass/fail data. Looking at cosmetic quality and deciding the standard is either good or bad.

The MSA technique is very similar to the one we've already looked at; in that we want to make sure we have consistency across difference appraisal systems.

Step 1 is to identify a set of standard pieces that will be assessed by each person or appraisal system that are usually assessing the product. Usually, 20 pieces are selected. This time though the samples are not randomly selected. The most important thing to remember is that you must make sure 50% are good product, 50% are defects. Any deviation from this 50:50 split will ruin the results.

In the table below 10 pieces are identified in the true standard as R for Rejects. 10 pieces are identified as A for Acceptable. Someone in your organisation who is considered to be the 'expert' on this quality standard has decided on these 20 decisions. This makes the MSA different in this case as there is a known or accepted true standard.

Each piece is then offered to an operator for them to make a judgement or offered to a camera system. If they get the judgement correct, and match with the 'expert' they score 1 if they get it wrong score 0

Part	True standard		Operator 1			Operator 2			Operator 3			
1	R		R	1		R	1		R	1		3
2	A		A	1		A	1		A	1		3
3	A		A	1		R	0		A	1		2
4	R		R	1		R	1		R	1		3
5	A		R	0		A	1		R	0		1
6	R		R	1		A	0		R	1		2
7	R		R	1		R	1		R	1		3
8	A		A	1		A	1		A	1		3
9	A		A	1		A	1		A	1		3
10	A		R	0		R	0		A	1		1
11	R		A	0		R	1		R	1		2
12	A		A	1		A	1		A	1		3
13	R		R	1		R	1		R	1		3
14	R		R	1		A	0		R	1		2
15	A		R	0		A	1		A	1		2
16	R		A	0		R	1		R	1		2
17	A		A	1		A	1		R	0		2
18	A		A	1		A	1		A	1		3
19	R		R	1		A	0		R	1		2
20	R		R	1		R	1		R	1		3
				15			15			18		48

This becomes like a little 20 question exam. Each inspection operator answers the 20 questions and then you work out their score. Operator 1 (System 1) in this case scored 15 correct out of 20 = 75%.

This is known as their **effectiveness**. It is the first of 4 statistics that will be worked out from the results.

1. **Effectiveness** – Individual and overall
2. **Probability of a False reject** – Individual and overall
3. **Probability of a False accept** – Individual and overall
4. **Bias** – Too tight or too loose with the standards – Individual and overall.

EFFECTIVENESS = $\dfrac{\text{No of Parts Identified Correctly......}}{\text{No of opportunities to identify Parts}}$

P (False reject) = $\dfrac{\text{No of times good parts are rated as bad}}{\text{No of Opportunities to rate Good Parts}}$

P (False accept) = $\dfrac{\text{No of times rejects parts are rated as Good}}{\text{No of Opportunities to rate Bad Parts}}$

Bias = $\dfrac{\text{P (False Reject)}}{\text{P (False Accept)}}$

The guidelines for these results are as follows:

Parameter	Acceptable	Marginal	Unacceptable
Effectiveness False Reject False Accept Bias	>.90 <.05 <.02 .8 – 1.20	.8 -.9 .05 -.10 .02 -.05 .5 -.8 or 1.2 – 1.5	<.8 >.10 >.05 <.5 or >1.5

Sometimes these guidelines look a little generous, but pass-fail inspection like this is always a poor way to grade parts. These guidelines are just a recognition of that fact.

The **Bias** result really should be greater than 1 to take this into account. You know that you are going make mistakes. If the Bias is above 1 it means that you're too tight with your standards and therefore the mistakes stay on your site and not passed on to the customer to find.

The output from SPC XL will look as follows:

	Inspection Capability		
	OP 1	OP 2	Total
Effectiveness	0.85	0.9	0.875
P(FR)	0.1	0.2	0.15
P(FA)	0.2	0	0.1
Bias	0.5	NA	1.5

Operator 1 has an effectiveness of 85% with a bias below 1. Which means they are too loose and therefore passing defects to customers? Although the overall result of 87.5% (.875) is Ok operator

1 needs some improvement and maybe some re-training. A good way to achieve this is to ask the best operator how they achieve their results. They often have simple informal rules. That allows them to be consistent.

'Oh, it's simple. If the mark is longer than my little fingernail, then I fail it'

Something like that. A rule, that you can teach to anyone and have an immediate impact. If there is no simple rule like that, make sure you collect a set of real boundary samples and put them on display at the point of activity. Your inspectors will never remember the correct standard otherwise.

If the system is camera based then settings will be important and light levels etc.

If you have any doubt about your measurement systems. Select the correct analysis technique and quantify the problem. In fact, it's unlikely that your measurement system is not adding to the problem that your project is trying to correct. The only issue is how much of a contribution is it making.

Conduct an MSA and find out.

Chapter 8

Special Topics for Measurement System Analysis

Do bad results always mean you have work to do to improve your measurement system?

We have given you the theory of how a basic MSA works for both Variable and attribute data, now we're going to add some experience to this approach to cover some special occasions when common sense should be used even though by the rules that we have set the measurement system fails.

Example 1

The Gage R & R fails but the result is too far inside the specifications to be considered problematic.

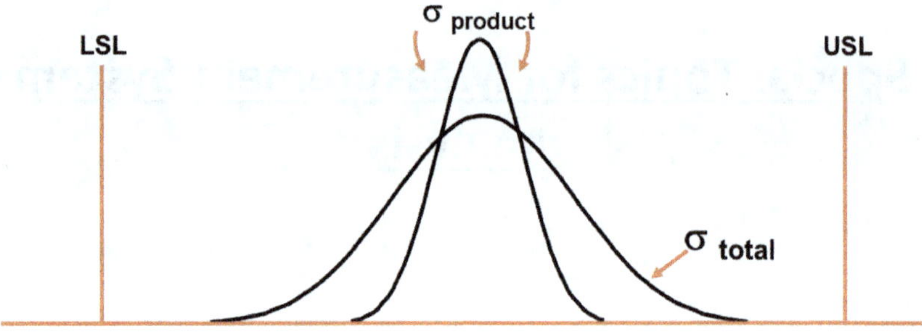

Although a result like this is theoretically a failed R&R result and the Measurement System should be worked on and improved. But is it causing any problems?? It is currently not over inflating the defect rate as there are no defects. And if the process starts to shift the measurement system is perfectly capable of identifying the shift in the process mean.

As you see in the diagram below it will trigger a slightly earlier signal that the process has moved and start to show rejects that are not currently there. But that is not really an incorrect signal the process is on the move and should certainly be audited to find the root cause.

If that audit is triggered by a defect that is not exactly present that is not really much of a problem. This early signal would be exactly what a process behaviour chart (SPC) would show if you were using one. This measurement system is happily controlling this process to a satisfactory level.

The risk therefore of using this measurement system is very low, this improvement work would not be a priority, use the measurement system with a note recording the MSA result.

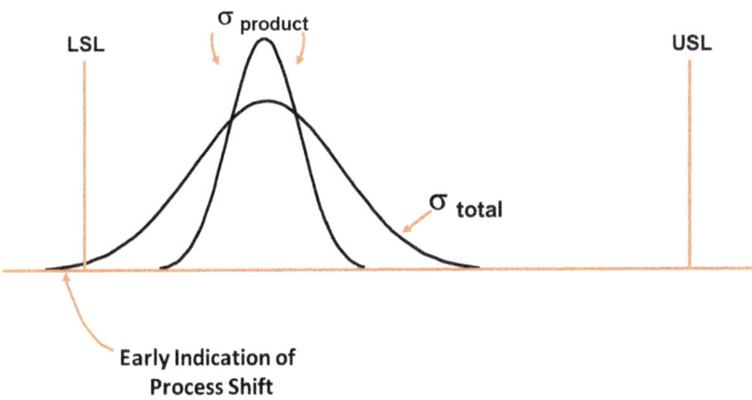

Example 2

The Gage R & R fails because the parts have very little natural variability.

This happens often in high precision processes but can happen in other processes too, remember the part of the R & R calculations are ratios compared to the natural product variability.

As we mentioned earlier the selection of the sample set used in the MSA is important because if you artificially select a group of parts that do not contain 'normal' process variability you can make the results look artificially bad.

When you make high precision parts where you are almost at the limit of the measurement systems precision it is not possible to follow the rules that the measurement system should have one more decimal place than the specification being used.

So, if your tolerance is to 100ths of mm it would be usual for the measurement system should measure to 1000ths mm, but when your tolerance is already down at 1000ths mm it is very rare for a practical piece of measurement equipment to be available at each machine capable of measuring 10000ths of mm. This is a lack of precision because of decimal places often means that in high precision work the measurement system could easily provide 50% of all the variability you're observing, and this would still be a piece of measurement equipment that you would use. In any other scenario your measurement system would be highly accurate and acceptable.

This poor result could also happen over time as you improve your process and remove natural component variation a measurement system that was perfectly capable, gradually over time the error from measurement system becomes more and more of the total variation that you see to the point where it is total contribution would be considered un-acceptable.

Example 3

This is not really an MSA at all, but an example of how simple tools can reveal measurement problems and or data problems. The tool in this example is the run chart. If you collect data, the minimum you should do once it is collected is to place it on a run chart and show it to the person in control of the process.

We have a case study video at the end of this paragraph to show the simple principle discussed here. The issue identified by the run chart is a lack of resolution of the measurement system. Not enough decimal places on your guessing stick. Your process is a random number generator, any pattern that indicates it's not a random pattern should be investigated.

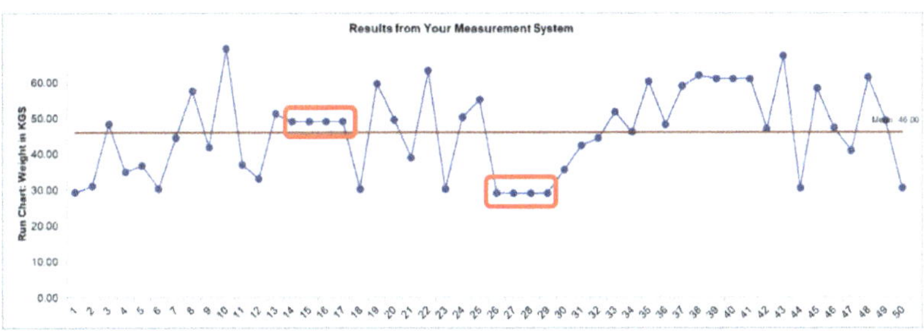

Does the data displayed in the above graph look like a random number generator? Random number generators don't repeat the exact result 4 times over. Something is wrong here; it doesn't necessarily have to be a measurement problem but it needs investigating.

Another deviation from a random number generator could look like this...

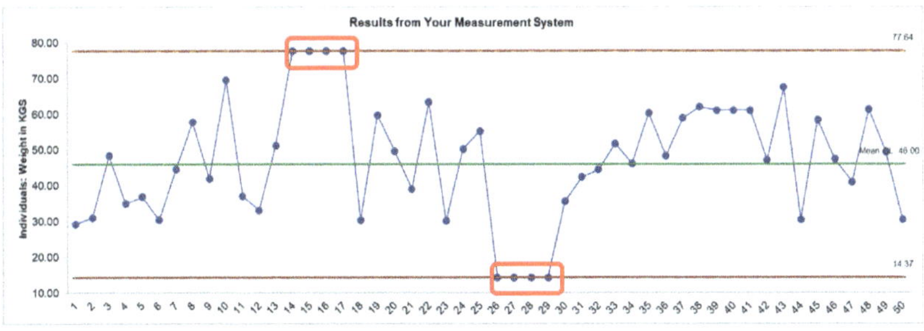

This process is displaying a random pattern most of the time, but when the results hit the upper or lower tolerance the process repeats exact results? That is strange, more than likely this is a result being corrupted by someone who does not want to report defect results. In this case the problem is unlikely to do with measurement system, but this is data you cannot reply upon.

That is the point of looking at your data, not to conduct an MSA in every case, but to be sure the data can be relied upon.

Chapter 9

Analyse

Starting at the beginning of your project the first question is your process in Chaos or Control? In other words, do I need to control the noise (Chaos – Standard Deviation) or do I need to move the Signal (Control - Mean). Controlling the noise will take 3 months and need a team. Moving the signal could take as little as 3 mins/3hours or 3 days.

For this you will need to use the diagrams we have already looked at

Run Chart
Histogram
Cpk chart

If you have pass/fail data, then only a run chart will be possible. Though the run chart will often tell you the answer immediately.....you can see it on the graph below, the dramatic period of chaos and the period of relative control.

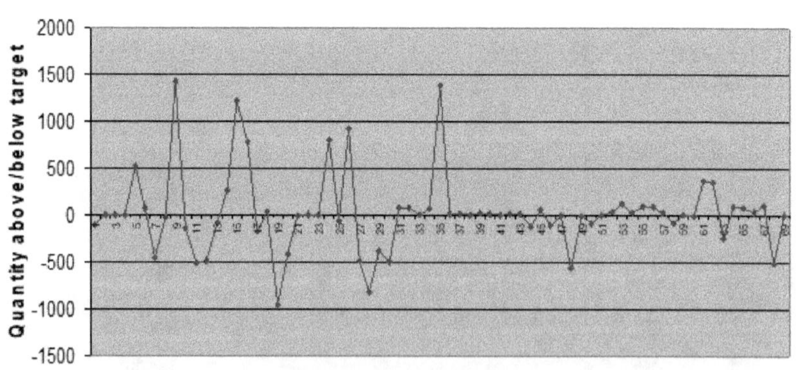

Essentially your process will be in one of 3 states....

CHAOS
CONTROL
EXCELLENCE

Most company's processes are in Chaos! And they are constantly striving to move them directly from Chaos to Excellence. This is not possible......

First step must be to move your processes from Chaos to Control and then from Control to Excellence. Actually what most people find is that once a process is in Control the problem that they have disappears, so they rarely progress to do the Excellence stage, they have too many other chaotic processes that they want to control first....

If you're doing any kind of lean implementation by the way, this is exactly the same approach. Lean will use standard work combination tables to create a standard way of carrying out a process and take the process from Chaos to Control.

So having used the data to decide the current state. Now we can analyse the process.

This is the point where the project really starts and where we re-introduce some tools that you were probably taught a long time ago but have never really found any value in using. Process flow diagram and a cause and effect diagram.

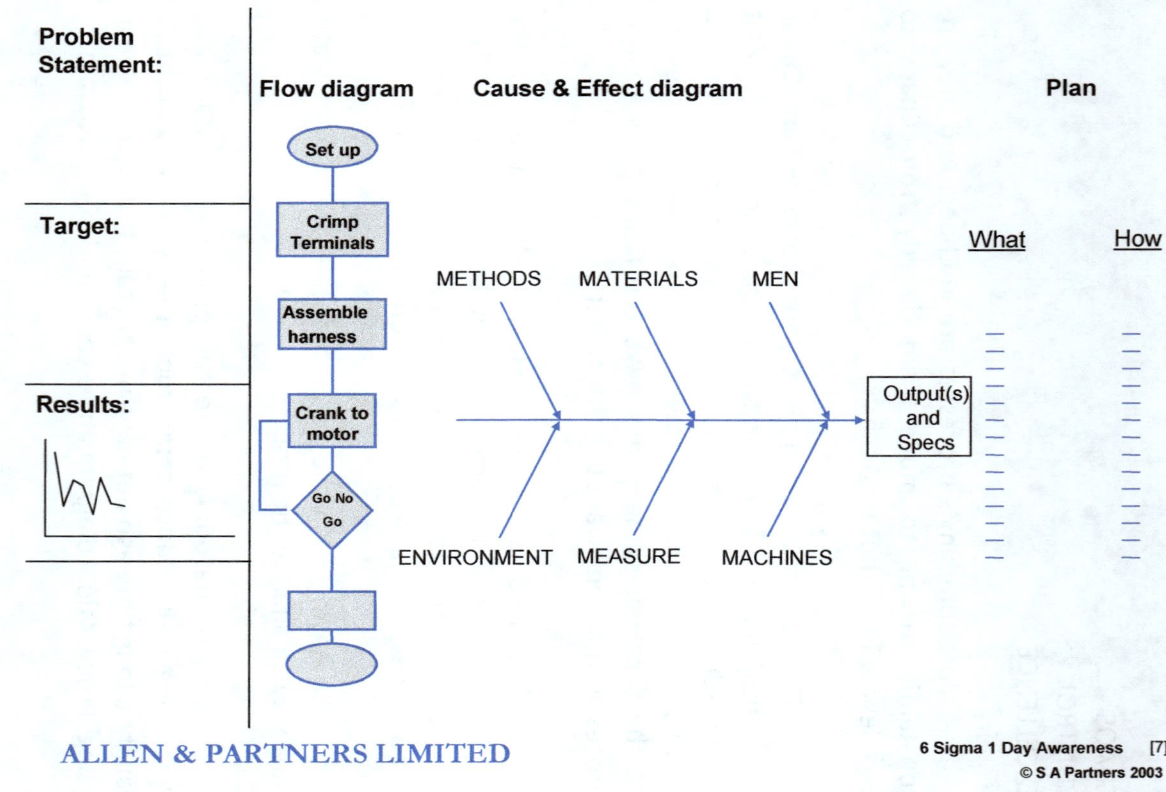

ALLEN & PARTNERS LIMITED

This simple analytical process we are going to go through is without doubt the singularly most important technique that I have learned since moving into manufacturing at 16 and was taught to me on my Blackbelt training by Air Academy Associates of Colorado Springs USA. It is known as PF/CE/NCX/SOP, I can tell you that the PF and CE stand for Process flow, Cause and effect, the rest I will describe later.

But first let me relate a story to show why this technique is so fundamental. At a relatively early stage in my engineering career I learned a new skill, capability analysis. Armed with a calculator and a data collection sheet pinched from Ford with the logo tipexed out (this seems to be an approach everyone used). I teamed up with a young QA engineer and we routinely took new press tooling and ran a capability study, this was the professional way to confirm the press tool was acceptable, so we thought. But when tool after tool demonstrated their incapability to achieving the specifications, like the histogram shows below we looked at one another and thought what the hell do we do now?

No one had ever taught us to squeeze the variability out of a process, only how to move the mean.

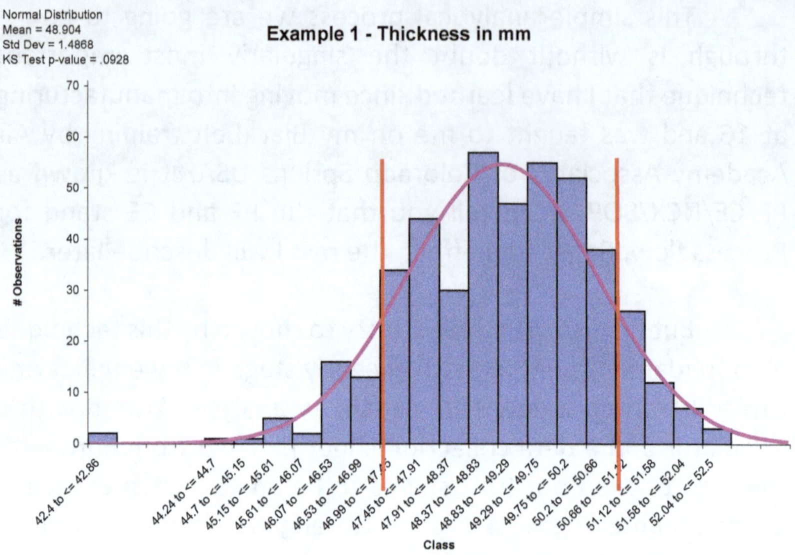

Think about it? When has anyone taught you to squeeze that distribution in to fit between the specs?

Like most engineers we knew how to move the average (the signal) of a process, that is pretty easy. But we had no idea how to affect the spread (standard deviation, the noise) and as most of our problems lay in this area we were pretty much stuffed. I never did learn what to do next until I was 35 years Old and was introduced to Process Flow/Cause and effect using the Statapult to demonstrate its effectiveness, still lots of things work well in theory in the class but crash and burn in the real world. But not this approach, the very first time I used it to bring a process under control it worked and I have been using it ever since.

It is a 'must do' on any project I undertake. It's fundamental because it does what we discussed earlier it identifies the inputs and allows you to fix them and when you fix them the process works! The process works because this is simple physics. If inputs never move, then your process will stay where you set it.

PF – Process Flow
CE – Cause and effect
SOP – Standard Operating Procedure

We looked at the process and the principles earlier in the text, now we shall add some extra advice on how to make this session a success.

In order to complete this, you are going to need the whole team for at least 1 day, you will lead the event do the writing and complete the diagrams etc and obviously get them turned into electronic records.
PF – Process flow

The flow diagram is a guide. It is not designed to replicate the exact logic of the process, you can tie yourself in knots trying to do that and waste a lot of time. But it should be as close as you can get to the logic in a reasonably quick time. It will be used as a guide to stop us shouting...

I know what it is!.... and jumping to cause.

It will make us evaluate every step in the process in relation to the failure effect that we are interested in controlling.

So normally if you get a team of engineers together and ask what is wrong, someone will shout out........*I know what is!......its the material!........No no claims another...it's the night shift! No don't be silly....we haven't set the winkie valve correctly!*

The flow diagram will help you to capture these points of view. But you will capture much more besides.

CE - Cause and Effect

This is when we will define all the process inputs and their current state of control. Using the flow diagram to guide you, visit every box on the flow diagram and ask yourself a simple question 'At this step what are the input variables that might contribute to the failure mode'. As they are identified write the variables on the cause and effect chart alongside the appropriate heading. Important points about this are:

1. Capture input variables **not** problems, if the team say 'but that's not a problem' remind them it's not about problems but inputs that vary or can be varied. We are simply looking for difference. Two setters using a different approach or machine settings are introducing variation. Both may make the process work effectively but to them it's not a problem BUT IT IS VARIATION! This part of the process is designed to capture real practice and not theory. Hence the reason why the operators and setters are the most important contributors. So don't let engineers and Supervisors over ride the input of the operators. For example, the operator states that the Air pressures varies by 15psi during a shift, but the engineer states this is isn't possible *'we bought a new compressor last year to correct this'*. One is theory, one is practice, and it's the practice that kills process capability not theory!

2. Getting the variable under the correct heading on the cause and effect diagram is not that important. The idea is that

when you have finished, all the machine variables might be given on block to a maintenance technician to control. The people variables to the team leader, material variables to purchasing etc. it is rarely as clean as that so a few in the wrong place is not going to matter too much.

3. Be as specific about the variable as you can, the more specific the easier it is to control or to investigate. An example would be saying that the steel coils vary in a pressing process. This is too general. If you specify steel width, length, thickness, hardness or % silicon now you can investigate further and if necessary improve/control it.

Now we have the variables on the chart you can add C, N or X against each of them.

C – Standing for **controlled** and means that the variable is under control. For that to be possible it must have a Standard Operating Procedure (SOP) that is:

1. Written down or displayed somewhere
2. You are happy with it
3. and finally you use it.

If you cannot meet **all 3** then it's not a C, any dial that is moved regularly by an operator or engineer for example is not a C!

Are you happy with your SOP's? Well an SOP should be clear and unambiguous. Here is an example.....in a cheesecake factory. They are trying to control the weight of the product to make sure it complies with the weight displayed on the packet and that they don't give too much extra material away. They start with the mixing of the biscuit base and the SOP says *'mix the biscuit until it is soft and crumbly'!*

What do you think? Yes, that SOP is plain nonsense! Check your Standard Operating Procedures. they are often like this....

Adjust the speed until you're happy....

Adjust the guide rails to suit the bottle...

A much better SOP would be....Mix the biscuit base for 150 seconds. It's clear and unambiguous. And now put a timer on the machine and fix it!

The best companies in the world know how to define a good SOP and stick to it without fail... no compromise. If you identify variables first, then make sure the controls are stated in the SOP. You're SOP's will be so much better. And please identify specific variables. I often see 'set up' as a variable on a Cause and effect diagram. Then the engineer says I'll write an SOP to control it. However, what they should have done is identified 14 variables that are in the set-up routine. Once you have that list of variables, you can write a great SOP that controls them all. Otherwise how would you do this?

So if it's not a C....then identify it with an...N

N – Stands for **Noise** or not controlled and is any variable that does not meet the C criteria above. Remember it doesn't have to be a problem to get an N just not held constant, day to day or hour by hour.

Finally...

X – Stands for experimental and is added to any variable that you want to find the true effect on the process. Later these X's will go into a designed experiment (D.O.E). X can be added to either a C or an N. So for example the press speed might be specified at 50 strokes per min. The standard is followed and well controlled. But is it that the right speed? Well this becomes an X and later we will find out if some other speed might be better.

When you have finished PF/CE/NCX/SOP/ISO9001 you have now defined the process, all the inputs and their current state of control. The whole process should have taken about 1 day to do.

But it is the most important day of the project. Nine out of 10 projects I've seen over the last 19 years, the process was out of control This includes working at Sony, where we thought we were brilliant at process control.

This technique exposes every possible area for improvement; it is a very bright light to shine on your process. Nothing else can be achieved until this analysis is completed as it creates a stable environment in which all other tools can be used. Into this stable environment other more mathematically analytical tools can be introduced and be successful, although many problems are fixed with just this technique. To complete this stage to control the variables usually takes 3 months, yes 3 months! I'd love to tell you that you can do it quicker but it's rarely possible. So if you're doing problem solving and think you can fix it in a week. If your process is in Chaos...think again!

If it is difficult to get your team together for one day you could split it in half, do a flow diagram in half day and the Cause Effect in half a day. If even that is too difficult the absolute minimum would be to get the team together for 2 hours to demonstrate this approach. This should take 2 hours approximately. Let them go back to work and then visit everyone individually at the point of activity and build up the flow diagram and Cause and Effect a job at a time.

When you have finished you should end up with something that looks like the diagram below. This is an example of a cause and effect diagram from a client I worked with some years ago.

One final tip. Maintenance is often the root of all variability. We often skip good maintenance and expect the technician to twiddle the knobs and make magic happen!

If the technician insists that they have to adjust the settings you have to remove the reason why they want to adjust and carrying out good repeatable maintenance will be a huge part of that.

CAUSE AND EFFECT FOR SANSOW ENCAPSULATION

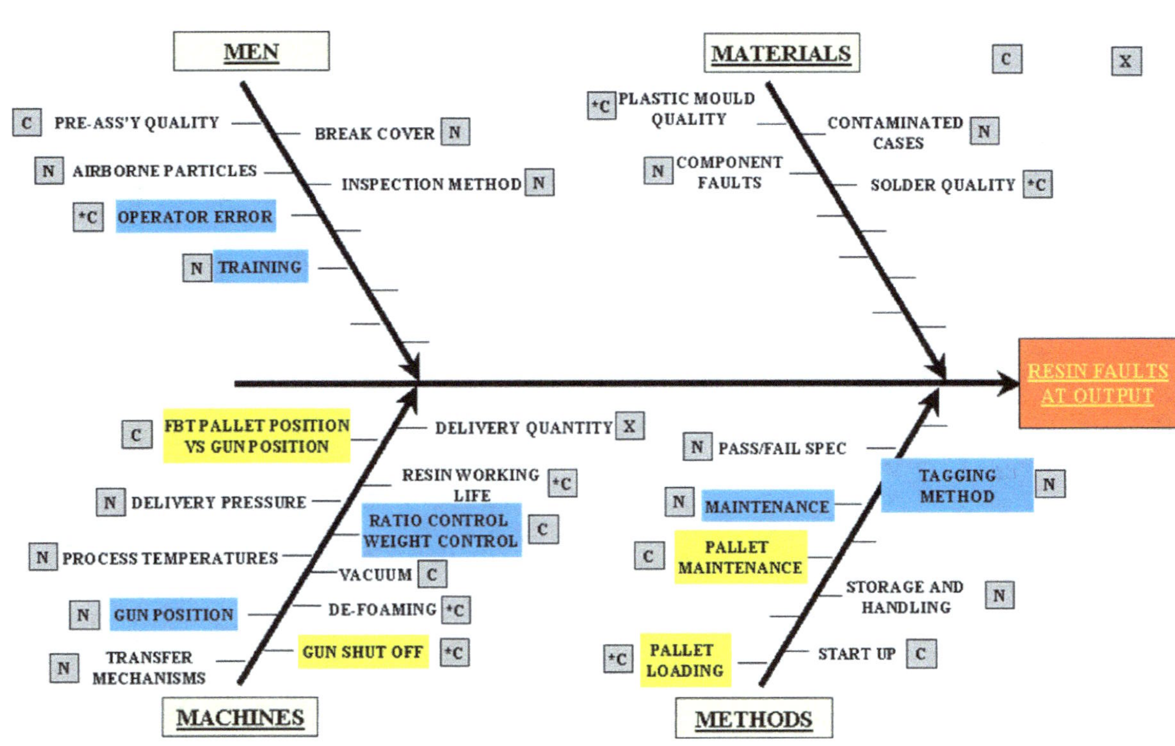

- The team will do this
- It will take one day – is everyone available? If one or two are missing it is probably OK, but the operator and setter must be present. The process DOERS must be present, not just engineers.
- Capture the practice not the theory, the operator sees the process cycle every day their view is most important.
- Flow diagram is a general description of what goes on
- Visit every box on the Diagram – What are the **VARIABLES** at this step that relate to my output? Avoid describing these as problems, we're interested in inputs that are not held consistent.

Everything is acceptable it is a brainstorming exercise at this point, even if you think it has a minor effect write it down anyway.

Write the variables on the chart avoid describing them as problems or solutions, just describe the variable, speed setting, cleaning routine, cleaning frequency, 3 shifts etc

At the end of this process we have a list of all the input variables and the current status of all the current controls. The aim now is turning as many N's into C's as we can, to reduce the natural variability of the process and make it more repeatable. It is not unusual for there to be upwards of 50 variables that are not controlled (N) you cannot control them all so what next?

A key point here is concentrating on the variables you can control and avoid moaning or concentrating on the variables you can't control. Also don't over analyse, you've decided that an input is moving, it shows variation. Don't try to analyse the benefit at this point, just fix it! and in practice most variables are controllable. They can be fixed.

Let me give you 2 examples....

A pineapple peeling machine, what variable did the technician moan about? The size of the pineapples! There's nothing we can do!

And because they took this view, they couldn't identify standard set up procedures for the machine....with a bit of clarity of thought however, they realised they could get the fruit segregated at the supplier based on size and they had 3 sizes...6, 7, and 8's

They then had standard setting for each size and bingo! We've got the process under control..

The second is ambient temperature of weaving thread; again, there is nothing we can do! We can't control outside temperature...

Instead with some thought they created a simple table, if the warehouse was at a particular temperature then condition the thread for a particular time before use. This made sure the thread reached the correct temp before being used. Simple!

In fact, most variables are controllable we choose not to control them, most of the time.

Chapter 10

Improve

OK now we're going to control all this noise....

This is going to draw in many more tools than just the quality we are introducing. But hopefully they are tools that you're already using in your organisation.

SOP – Good standard Operating Standards
MSA – Measurement System Analysis (Ask your quality department to run one)
5s – Workplace Organisation (Stop using 5s to clean up and use it to control Processes)
Poka-Yoke – Mistake proofing, automatic alarms, jigs etc
TPM – Total Productive Maintenance (Controlling Machine Variability)

Noise control

Being faced with 50-200 variables is often were the project starts to go off the rails. Six Sigma courses can be packed with lots of analytical and data mining techniques and at this point your experience is such that you probably don't know which one to use. Controlling N's is the aim, so the analytical approach should help us priorities or focus, although not all the approaches are statistical in nature here are some ways to get started on process control improvement.

1. A technically simple problem - Put a control in for every uncontrolled variable (N). An example of this would be improving set up time. As most of the variables are about distance, location, organisation or specifying machines settings, using tools like 5s and TPM. All of these variables can be controlled. In fact, it's only by controlling them all, that you get the

combined effect and reduce the set up time to the practical minimum very quickly.

2. A technical problem with obvious groups of variables – This approach is about finding solutions themes and then implementing the controls in one mini project. An example would be machine settings, say there are 10 machine (knob settings) settings, speed, feed rate, pressure etc which are not agreed as standard by the technicians. These would be grouped together and fixed at the same time in one meeting with the technicians. So one piece of work fixed 10 variables, it's great way to work and get your process under control fast. Examples of solution themes would be:

    a. Machine settings
    b. 5s controllable variables
    c. Supplied materials variation
    d. TPM controllable variables (Maintenance)

   These solution themes often become apparent during the process of constructing the CE diagram. However, if you need to, an affinity diagram might be needed to gain agreement with the group if they are not obvious. But as you become expert at the analysis you should be able to this without the need for an affinity diagram.

Variable Controls Planning

Variable	Priority Or X	MSA	SOP	Poka yoke	5S	Special Action	5 Y's	TPM	Controlling process

OK now you've defined the process, all the inputs and their current state of control and you have decided were to start.

Above is a simple document to help you plan and track your progress. It will also make sure that the fix you implement really is sustainable, true control. The Variables control plan above, shows you the list of improve tools you can potentially use.

MSA – Measurement System Analysis
SOP – Standard Operating Procedure
Poka-Yoke – Mistake Proofing -
5s – Workplace Organisation
Special Action – A one off fix it type action
5Y's – Be more specific with the Variable. Don't just say the material is variable. Specify, it's the thickness or the density etc. Then you can find a solution
TPM. - Total Productive Maintenance (Machine variables)

These improve tools are there to design a specific control technique. This design of controls is actually the improve stage of the project. True control will come when you attach an audit or check or maintenance procedure to lock these controls in..

The document above simply has a space to list all the variables and then you can simply tick the tool you will use to control that variable. Are you doing an MSA, writing an SOP, using a mistake proofing device, 5s/workplace organisation, taking a special action to fix something or using TPM or a combination of these techniques.

As you can see this were the Lean tools begin to come in. 5s and TPM are methods for controlling variation, so is Set up reduction or SMED as it often called. One key point to remember is that variation often breeds variation. Process settings often cannot have fixed standards because maintenance is never done regularly. They have to constantly come up with 'new' settings to get around the variability of the

Machine condition. Fix the maintenance then fix the standard process settings.

However, of all the columns on the variable control plan the most important column is the CONTROLLING PROCESS.

Without this phase the improvements will just deteriorate, your process loves to be in CHAOS. It will just slide back into that state over time. Control is covered in the next chapter.

Chapter 11

Control – Real Control

This is the method that ensures the Improvement technique that you've chosen and specified is adhered to. Being in a state of control is not a natural state for a process and it will not happen by accident. Control is essentially a policing process and is the bit that in the past I was particularly poor at. Consequently, my project improvements would slowly deteriorate. The controlling process is the process that makes you adhere to the SOP's and other rules that you put in place in the Improve phase.

Let's say for example the Standard is that the filters are cleaned once per day on a process, a simple thing to decide, but what is the controlling process?

How are you going to make sure this done? Well there are simple techniques to do this that you already use.

- A start up audit with a sign off sheet might be one way of doing it.
- You might visualise the check, like they do when cleaning a rest room
- A TPM check.
- An ISO 9001 Audit
- A 5s Audit

If some of these techniques are currently viewed as of little value and a chore, then you really should re-assess why to do them. ISO 9001 should control your process. 5s should control your process. If that is what they are doing, then you would value them. If ISO 9001 is just about getting a certificate, then it is a burden and simply waste! Build the control of input variables into your ISO audit, build them into your 5s audit and make them valuable. Better still conduct one process audit that you use for ISO, 5s, safety, NADCAP, FDA etc. Then when you are assessed, you are assessed on behaviour that is second nature to provide great products to your customers, not to get a certificate that means nothing to your actual customer service and performance.

All these audits and checks should be visual and displayed in the area, so that as management walk through, they can informally check that the status is green. This might be audited every week by the quality department just to make sure it is done. But it is vitally important that management pick up violations of the rules and demonstrate how important these controls are otherwise why would people do them if the management don't care whether they are followed.

THERE IS NO BETTER WAY TO ENSURE CONTROL THAN THROUGH SENIOR MANAGEMENT BEHAVIOUR.

It is not something to cover in this text but Leadership Standard Work is crucial to create a culture of control and process improvement.

In fact, management discipline is one the most important but informal controlling process you can use. A manager should never walk past a shadow board with a tool missing and not challenge it…a shadow board is a Standard that should be adhered to. They should never walk past audits that are out of date and not challenge it. If they do, it's a clear signal that standards do not matter and everything will slowly deteriorate.

In the best companies in the world, management discipline like this is second nature. At Sony for instance, the more senior the manager the closer they would look at the process controls. Visits from Senior Japanese Managers where a nightmare, they would look so closely at the process. They would nearly always find something out of control! Our processes never slipped back to CHAOS however...and if I was having a great day producing TV's and dead on target. My M.D. would appear and rarely congratulate me on the output. He would always go and audit my controls and find something that has slipped that he wanted correcting. Western Management rarely behave like this.

Chapter 12

Statistical Process Control (S.P.C)

Walter Shewhart developed Statistical Process Control (SPC) at the bell telephone laboratories in the 1920's. Shewhart developed the control chart in 1924 and from that point on everything that Modern world class quality control is based upon is the idea of understanding and reducing variability.

We have already looked at using a simple run chart as a primary measure. But now we're going to enhance that run chart and turn into a control chart. We've waited until now so that we have the process under control. The enhanced chart we can use to pick up process deterioration signals within a few minutes or hours of them happening and help us to diagnose and correct those problems.

Of all the things that I see when I work for various companies, SPC is without doubt one of the most mis-used, under used or misunderstood tool. In most cases because it is mis-used, it is pointless and creates waste and inefficiency. Some of my clients have a saying on the shop floor.

Parts not charts!

For those of you that use SPC and for those of you that aren't. What is Statistical Process Control? Amazingly when I ask this question to companies that use it, most do not really know what it is.

A simple Phrase that is often used is to 'Listening to the voice of the process' and it certainly does that. Another phrase that I think is great is checking the health of the process. What it isn't and must never be is checking the health of the parts. That is comparing parts against specifications and is often done on these charts. That is just product grading. But that is wrong.

SPC is checking the natural behaviour of the process and when that behaviour becomes un-natural it will create a signal to investigate. A PROCESS BEHAVIOUR CHART....

Why use S.P.C - https://youtu.be/3FdFSAsRGoE

Consider the situation of your manufacturing process. All processes are essentially big random number generators. Of course, that random number is defined by all those variables that we've been discussing so far in this book. Now one of the problems with a random number generator is that you don't know exactly what's coming next and that makes a process difficult to understand and predict.

Let's consider a random number generator that is relatively easy to understand, easier than a manufacturing process anyway....A dice. A Six-sided dice.

It's clearly a random number generator. We definitely don't know what the next number will be. But what do we know about this process?

Well, essentially, we know 3 things that can help us understand the health of the process...

1. We know the numbers will fall in a boundary of 1-6. This is the spread or range of the process. (The Noise)
2. We know that all the numbers have an equal chance of occurring. The distribution of results is even. (The shape of the data set)
3. We know the average, which is 3.5 for a dice (The Signal)

So we know....

The Signal - The Middle or average
The Noise – The spread or range (Standard deviation possibly)
The Shape – Or distribution of the data

If I rolled a dice on the table right now and told you the result was 7!

What would you think? That's not right? That's odd?

We have broken rule number 1 above, the spread....

Now at this point you don't know what is wrong but you know that this is a result we do not expect from this process and we should investigate what's wrong. Your process is not healthy!

How about if I keep rolling 2's.......again it is an odd result. Not what we are expecting from this process. It would not follow the distribution we expect and, if it keeps happening the average of the process will be reduced. Not be 3.5....so we would investigate, what is wrong with the process?

And that is pretty much SPC. If we can understand the spread, the shape and the average for any process then we can understand when it starts to do something unusual. We should then investigate and correct the problem. All we are really doing is understanding what is normal for this process. When the process is doing something that is not normal, we investigate.

Now in my example of rolling a dice what wasn't mentioned?

Specifications!

Specifications have no place on a control chart and should not be used to indicate if you have a problem with the process.

Specifications are used to judge the product....they are not used to judge the process.

Control limits are used to judge the process....

Specifications are also used to judge process capability...or the health of the parts.

Control limits are provided by the process itself. They are the **voice of the process**. Because they are provided by the process, they are natural to the process. Since the process decides what they are. After all, where did the limits of 1 and 6 come from for the dice? They came from the dice itself. Had my dice been 10 sided the limits would have been 0 and 9....

Using data from your process you can decide what the natural limits are..BUT

Here is what happens if you use specs instead to control a process...

Lets assume that I've rolled a dice 1000 times and it has consistently produced numbers from 1 to 6 in the same quantities with an average of 3.5..

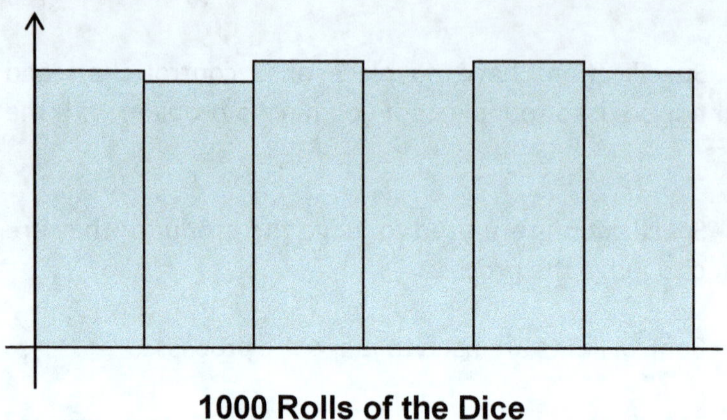

1000 Rolls of the Dice

We would be perfectly happy that it is working correctly. Now suppose we get a phone call. A company has heard about our process and wants to buy some of those numbers. But....they tell me that they don't like 1's and they don't like 6's. Please don't send any of those. So their specifications are....2 and 5.

When we turn our on process on at the start of the shift, what is our locked in guaranteed defect rate?

1/3 or 33%.. A defect rate the operator cannot improve upon. This is going to be the best they can do.

OK if you look below we've now imposed the spec limits on our control chart.

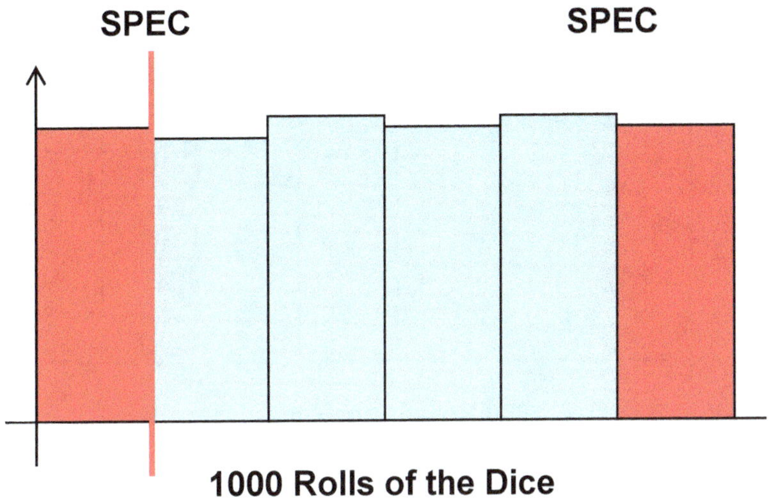

We start to run and initially everything is between the specs....then we roll a 1...a defect!

Now when a technician see's a defect what do they want to do? Adjust the machine to eliminate it of course.

OK, how much do we adjust the process? Shall we make the process 1 bigger? Adjust it to the right...CLUNK.....now that means we'll no longer get 1's...but of course we can get 7's now...

On we go, more good product indside the tolerances, then we get a 7! More defects!....we'd better adjust....let's move the process 2 lower, 2 to the left on our chart. CLUNK! CLUNK! This means that we can now get 0's

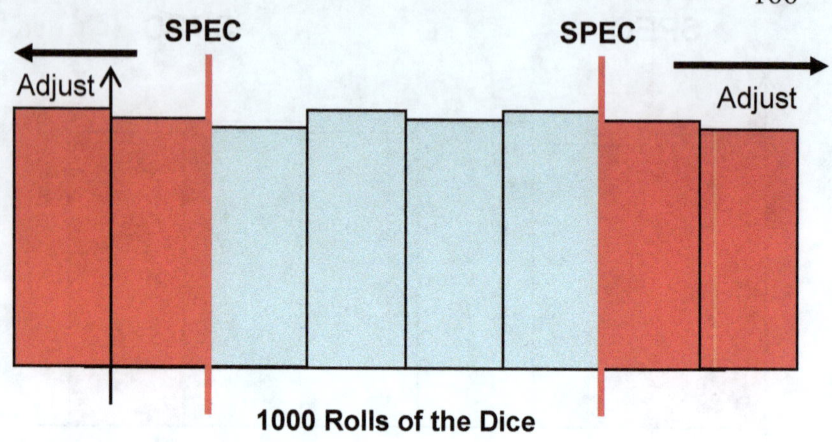

Look at the chart we are creating. When we arrived at work we had a 33% defect rate. That was the best we could achieve. Now that we've been keeping an eye on it and using the specs to adjust we have a.....50% defect rate!!

This adjustment is known as over reaction or over control. Over control always results in increases in the defect rates. The say that 90% of the time process adjustment is unnecessary and is making the process worse! SPC reduces over control to a minimum by helping to make the operator or technicians responsibilities clear. They should hit the target (land the average on the nominal value) and they should keep the process stable, but they cannot effect the natural limits of the process if they try, they will make the spread worse not better. They should not refer to specifications when controlling...when the process is on target and stable, they should drink tea, read the paper and count the cash and keep hands off!

Let's think practically what is the best strategy for a technician to minimise the process defect rate.

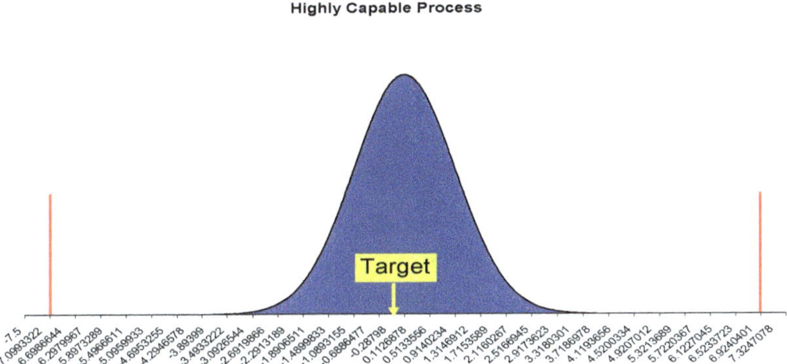

Look at the process above where is the best place for this process to be set. Hit the target...hit the middle of the spec

Look at the process below a very incapable process. What is the best place to set this to minimise the defect rate? Hit the target, hit the middle of the spec.....at no time did we need the upper and lower limit of the spec to minimise the defects....we only need to hit the middle of the tolerance. The nominal.

If they do that they will keep the process STABLE...

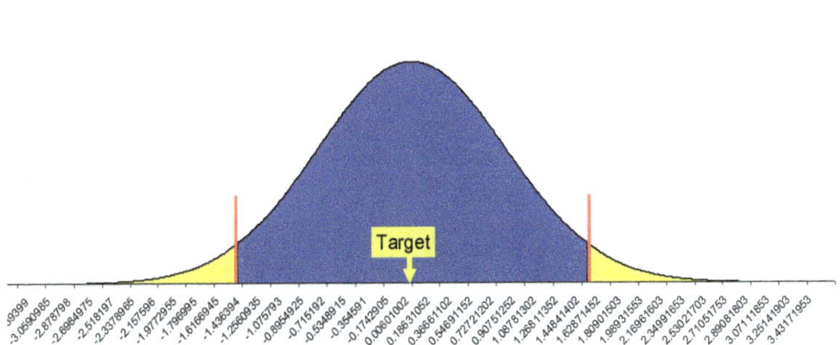

Both of these processes could be said to be in control. The text book definition of a process in control is STABLE and PREDICTABLE

STABLE – The mean does not move

PREDICTABLE – the Standard deviation or spread of the Data does not change.

And to achieve that, the best thing an operator can do, is put their hands in their pockets, drink tea and read the paper! In other words, stop playing with the inputs! You'll only make it worse!

The natural limits reflect that natural spread and prevent over reaction.

All the spec does is check the health of the parts good or bad. The operator can grade the parts good and bad. But they must not use the specs to grade the process. If they do use the specs to adjust the process above it will often result in more defects.

Now of course when you look at these diagrams, most people would intuitively know what to do. However, this is not what the operator or technician see's. No one presents these single diagrams to the operator. What the operator sees are single data points that they judge pass/fail using specs and then react!....

Let me show you...

The operator sees the single data point in process A, what should they do?

Well the process is perfectly centred, there is nothing they can do to improve the situation. They should carry on drinking tea and reading the paper! To adjust this process will only make it worse. Only the management can fix this defect rate. This problem is about controlling noise or common cause variability. It needs a project and it will take 3 months to complete using the tools we have discussed earlier.

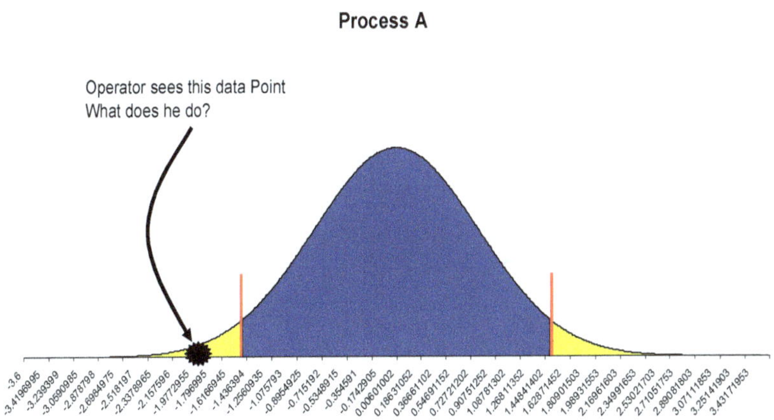

Now look at Process B below. The operator see's the exact same data point. Now what should they do? They should take action, not necessarily a process adjustment by the way. But they should investigate why the process is off target and take appropriate action to centre it. Essentially the operator is really in charge of moving the aim of the process and nothing else. Most dials and adjustments on a process move the aim.

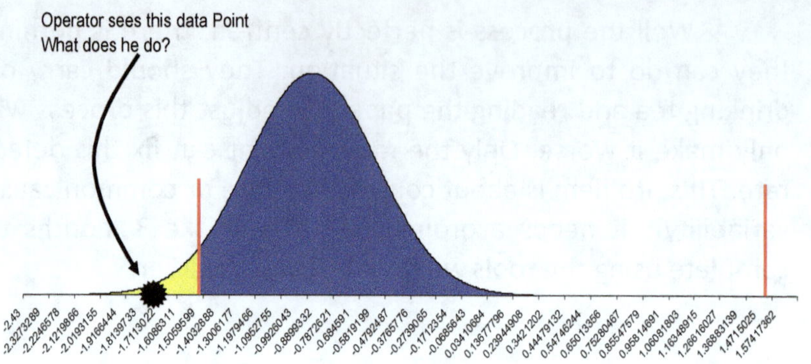

You see the operator has no clue where the centre of his process sits when he looks at a single data point like this....pass/fail. Then they make an adjustment, sometimes correctly, sometime incorrectly...however the senior management are responsible for fixing Process A defects. It's a capability problem. The technician is responsible for fixing process B defects. It's a control problem. But neither the management nor the technician is aware of this!

Everyone thinks the operator or technician is responsible for the defect rate **always**!

The control chart is designed to show the operator or technician the natural boundaries of these process distributions. It also shows the management; which medium term projects are needed with teams to run them.

Another way to look at these 2 problems, the technician or operator is responsible for the mean, the signal or the centre of the process. The Management are responsible for the noise or the spread of the process. This is the Process physics that we introduced in Chapter 2. Without understanding the process physics no one knows what is the correct action to take.

When these 2 groups understand their responsibilities this is when low defect rates are possible. To think that the operators are solely responsible for the defect rate is madness and will actually make sure your defect rates are much higher than they need to be. I've even heard some companies basing the company bonus scheme on the defect rate. Utter madness, the Senior management are responsible for the defects not the local or process management.

Process control can be defined when a process is stable and predictable. This is what the technician is responsible for. That does not mean defect free. To be defect free the process they have been given to run, must also be capable.

Capability is the responsibility of senior management. If the senior management provide a process that's not capable and expect the technician to deal with that. There is only one thing the technician can do...make it worse!

So all we need to be able to understand any process, is the middle, the spread and the shape. This will tell us how the process normally performs and we can use this information to tell us when something has changed in the process and stops us over controlling and making the defect rate worse.

Unfortunately, unlike a dice there are no obvious hard limits for your process results. So if we are going to create a control chart where do we set our natural limits? First we have to collect data usually between 50 and 100 data points.

We need limits between which the process will fall most of the time, a natural region. Limits, that minimise the chance that we get a false signal of a process change and over control the process. But also minimise the chance that we will miss a signal if one is present and fail to act when we should.

When Shewhart looked at this the most practical place for limits that performed the best and minimised these two types of error was 3 standard deviations from the midpoint of the process. If you try to consider this from a statistical or mathematical point of view (Something Shewhart did not do). The chart below shows that if the normal distribution is at play then you would expect 99% of your results to be in the +/-3 sigma boundary.

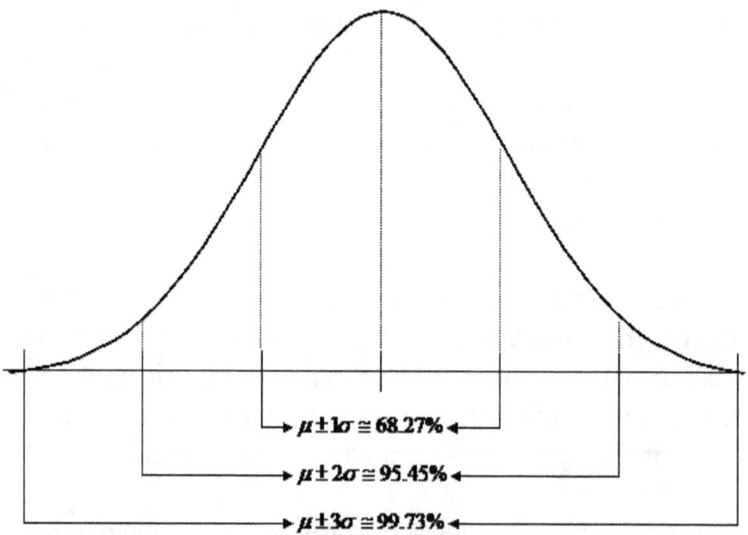

Shewhart simply used empirical evidence to see where the limits worked best. But this diagram might help to describe in part why the +/- 3 sigma limits work so well.

So the theory is to go out to the process collect data. Calculate the mean and the Standard deviation. Go up 3 x standard deviations and down 3 x standard deviations from the mean and this is where the control limits will be placed. That would require the following formula to be used for Standard Deviation..

$$s = \sqrt{s^2} = \sqrt{\frac{\sum_{i=1}^{n}(x_i - \bar{x})^2}{n-1}}$$

At least that's the theory anyway. In practice Shewhart came up with slightly different way for calculating the limits. To understand why he did this, think only of the fact that it was 1924. He had no calculators or computers. He wanted to make the calculations as simple as possible. Using mostly the 4 basic arithmetic operations. Avoiding the square root operations in the standard Deviation calculation above.

The calculations below are for the 7 Diagrams that Shewhart introduced and the first 3 calculations can be done pretty much by hand, without the need for a calculator, or computer. One that also isolates genuine machine influences from other influences that are external to the machine. So that it doesn't matter how the data is collected it will always isolate them.

These calculations are covered in the table below.

	Chart	CL	UCL	LCL	Comment s
1	$\bar{x}-R$	$\bar{\bar{x}}$ $\bar{R}$	$\bar{\bar{x}}-A_2\bar{R}$ $D_4\bar{R}$	$\bar{\bar{x}}+A_2\bar{R}$ $D_3\bar{R}$	$\hat{\sigma}=\frac{\bar{R}}{d_2}$ Use when n < 10
2	Individuals with Moving range	$\bar{\bar{x}}$ $\bar{R}$	$\bar{\bar{x}}+E_2\bar{R}$ $D_4\bar{R}$	$\bar{\bar{x}}-E_2\bar{R}$ $D_3\bar{R}$	$\hat{\sigma}=\frac{\bar{R}}{d_2}$
3	$\bar{x}-s$	$\bar{\bar{x}}$ $\bar{s}$	$\bar{\bar{x}}+A_3\bar{S}$ $B_4\bar{s}$	$\bar{\bar{x}}-A_3\bar{S}$ $B_3\bar{s}$	$\hat{\sigma}=\bar{s}$ Use when n>10 or when n varies
4	np	$n\bar{p}$	$n\bar{p}+3\sqrt{n\bar{p}(1-\bar{p})}$	$n\bar{p}-3\sqrt{n\bar{p}(1-\bar{p})}$	n is a fixed size
5	P Proportion	$\bar{p}$	$\bar{p}+3\sqrt{\frac{\bar{p}(1-\bar{p})}{n}}$	$\bar{p}-3\sqrt{\frac{\bar{p}(1-\bar{p})}{n}}$	Use n_i instead of n_i if n 's vary widely
6	C Count	$\bar{c}$	$\bar{c}+3\sqrt{\bar{c}}$	$\bar{c}-3\sqrt{\bar{c}}$	Fixed area of observation
7	u	$\bar{u}$	$\bar{u}+3\sqrt{\frac{\bar{u}}{\bar{a}}}$	$\bar{u}-3\sqrt{\frac{\bar{u}}{\bar{a}}}$	Use a instead of a if a 's vary widely

I'm not sure they look like it but this is the simple way to calculate limits. The complicated way would be to calculate Standard Deviation long hand using the formula.

The reason the calculations in Shewharts table look a little complex is mainly because these are all the calculations for 7 different types of SPC chart developed by Shewhart. They also contain the use of a table of constants shown below. These constants essentially allow for the range calculation to be used instead of Standard deviation. The constant 'converts' range into an equivalent standard deviation value. The table of constants is shown below.

n	A_2	A_3	B_3	B_4	d_2	D_3	D_4	E_2
2	1.88	2.66	.00	3.27	1.13	.00	3.27	2.66
3	1.02	1.95	.00	2.57	1.69	.00	2.57	1.77
4	.73	1.63	.00	2.27	2.06	.00	2.28	1.46
5	.58	1.43	.00	2.09	2.33	.00	2.11	1.18
6	.48	1.29	.03	1.97	2.53	.00	2.00	1.11
7	.42	1.18	.12	1.88	2.70	.08	1.92	1.05
8	.37	1.10	.19	1.82	2.85	.14	1.86	1.01
9	.34	1.03	.24	1.76	2.97	.18	1.82	.98
10	.31	.98	.28	1.72	3.08	.22	1.78	
11	.29	.93	.32	1.68	3.17	.26	1.74	
12	.27	.89	.35	1.65	3.26	.28	1.72	
13	.25	.85	.38	1.62	3.34	.31	1.69	
14	.24	.82	.41	1.59	3.41	.33	1.67	
15	.22	.79	.43	1.57	3.47	.35	1.65	
16	.21	.76	.45	1.55	3.53	.36	1.64	
17	.20	.74	.47	1.53	3.59	.38	1.62	
18	.19	.72	.48	1.52	3.64	.39	1.61	
19	.19	.70	.50	1.50	3.69	.40	1.60	
20	.18	.68	.51	1.49	3.74	.42	1.59	

I'm not going to go through every calculation for every type of control chart. But there is a video taking you through the creation of an X Bar R chart.

My advice for what it's worth. Get your software to make the calculations and use the chart from there on....

Here is an example using Mintab to create an X bar R chart that you might find helpful and an SPC XL example follows it..

https://www.youtube.com/watch?v=WBgR0rpSn6I (X bar R chart)

and a Moving Range chart created using SPC XL https://youtu.be/6fW2JTgavAY (Individual Moving Range ImR)

These calculations for all the Control Charts put limits at essentially +/- 3 sigma for all the different types of data sets that could be collected using Shewhart's methods.

1. Sampling a sub-group less than 10.
2. Sampling individual points only.
3. Sampling subgroup greater than 10.
4. Counting Rejects from a constant sample size.
5. Counting rejects when the sample size changes.
6. Counting flaws on the same item or assembly.
7. Counting flaws on different items or assemblies.

So you've identified the fact that you need control charts to stop you making the process worse. You've correctly identified the chart to use. Collected some data usually 30 - 50 data points worked out limits and you're ready to use the chart to spot unusual trends. Trends, that don't match the performance of the original pattern that the process displayed. Essentially these trends are non-random patterns that you wouldn't expect your random number generator (your process) to produce.

To standardise this process a number of out of control rules have been defined to help us to see what is 'not usual'. These rules where not originally defined by Shewhart, who liked to let the operator investigate any pattern that didn't look right. But where added later.

They have become known as the 'Western Electric rules' and were agreed in 1956 to ensure a uniform way of interpreting control charts by engineers and line workers.

1. One or more points are outside the limits – this is like a dice rolling 7!
2. Seven Consecutive points above or below the average – Like going to the casino and getting 7 Reds in a row at Roulette, not impossible but unlikely unless something has happened to the process.
3. Seven consecutive points are either increasing or decreasing – Like watching the lottery balls come out in order highest to lowest.
4. 2 out of 3 points fall more than 2 sigma from the mean on the same side. Only 2% of our data should be in this area of the graph but we are seeing 66% of our data here, this is likely constantly rolling 2 on a dice. Not expected.
5. 4 out 5 points fall more than 1 sigma from the mean on the same side. Only 16% of our data should fall in this area. We are seeing 80% of our data in this area, this isn't expected.
6. 14 consecutive points alternate up and down repeatedly. Often this is a sign that you have 2 of something in your process. 2 Impressions in a mould tool. 2 machines feeding one machine. 2 technicians using different settings.
7. 14 consecutive points are all 1 sigma from the mean – now if this is true, then apart from finding out why and locking in the change. The next thing you would do is re-calculate your limits. Something that you would never

do with specification limits. As now 100% of your data is in a zone where only 68% is expected.

It's worth noting that Shewhart used rule 1 and rule 5 and let the operator investigate any other pattern that didn't seem correct.

Ok you've selected the correct chart and understand the out of control symptoms rules. Let's look at setting a chart up..

Step 1. – Get the process in control and establish rules for all the variables (Process Flow, Cause & Effect NCX/SOP)

Step 2. – Collect Data to work out the trial or original control limits. Using some suitable software, calculate the limits. Below is an example. It's an Individual moving range Chart for the measurement of Noise in decibels

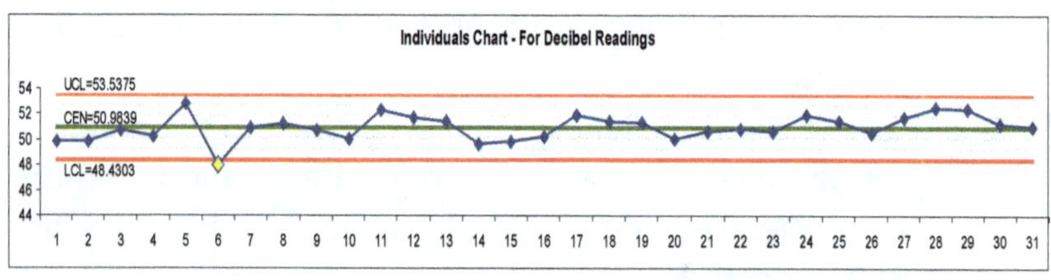

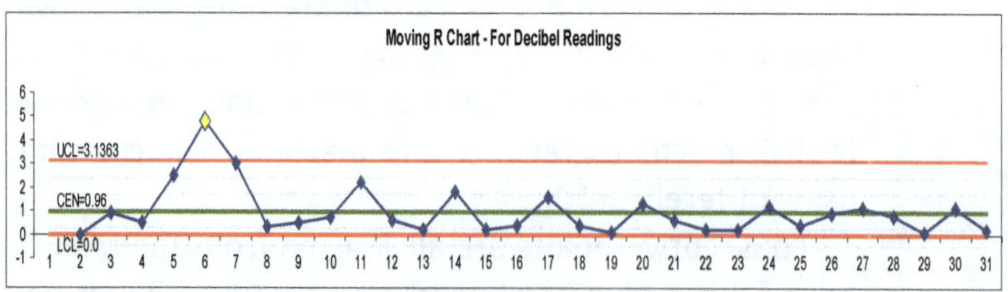

Step 3. – Remove any out of control points from the calculations. In the sample above, data point 6 is shown by the software to be out of control. This data point must be removed and the calculations re-done.

Step 4. – Extend the New limits across the whole chart and use at the point of activity to find and eliminate out of control causes.

Step 5. – On a regular basis, re-new the chart. Calculate the limits again and issue a new chart. At the same time calculate the Cpk for this process.

Ok you've selected the correct chart, set it up and you know what the symptoms are and that they indicate something might have changed. Now what should the operator do next?

This part is probably more important than the chart itself. Because now this is an indication that the operator is allowed to adjust the machine, right?

Wrong!!

Before you even consider using a control chart you should have gotten the process under control and specified all the controls for all the variables. The control chart really only monitors control it doesn't give you control. If it shows an out of control signal that tells you a control is not being adhered to, the operators' job is to find the rule being ignored and re-set back on standard.

Therefore, the operator is allowed to audit the process controls and if one is not correctly set. They are allowed to put it back to the correct setting. If all the inputs are correctly set.

Then they should step away from the machine and call for help! They do not adjust away from standard. Look at the example below....

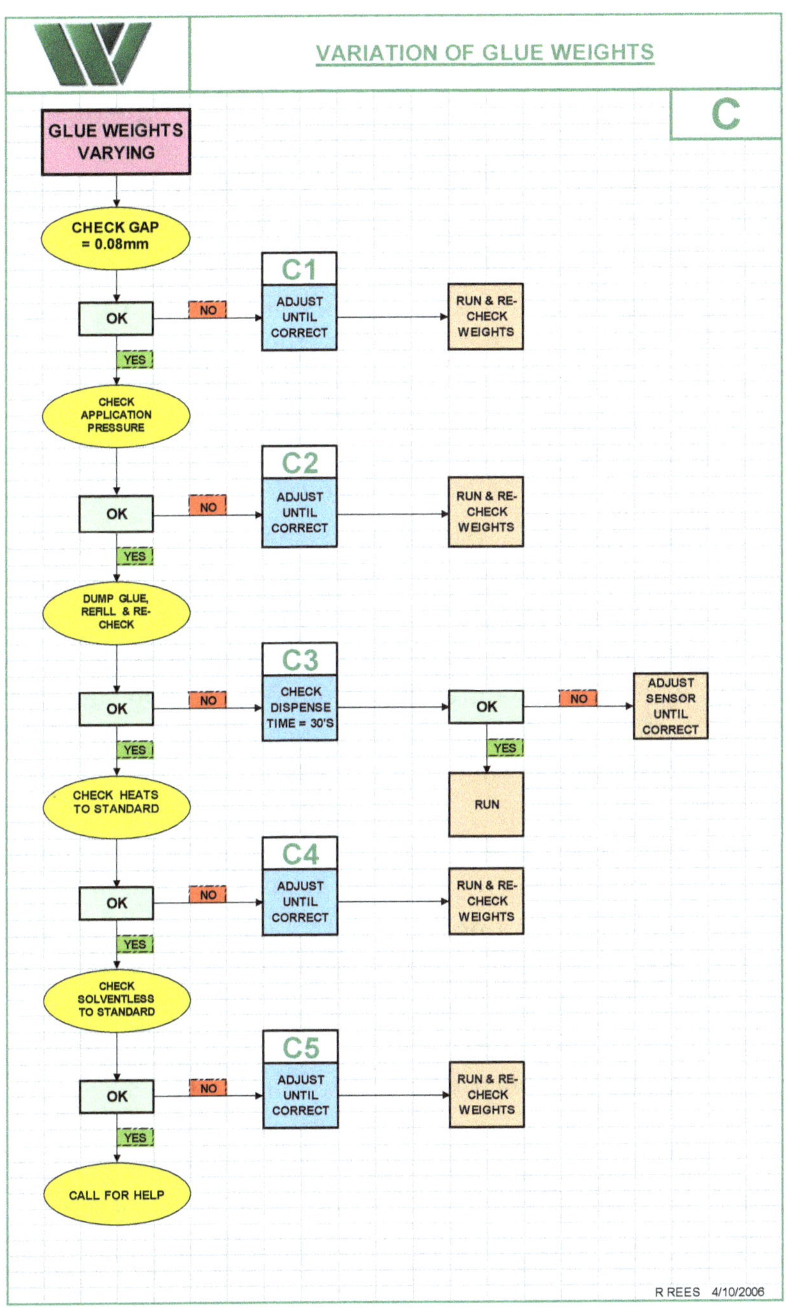
VARIATION OF GLUE WEIGHTS
C
GLUE WEIGHTS VARYING
CHECK GAP = 0.08mm
OK
NO
C1
ADJUST UNTIL CORRECT
RUN & RE-CHECK WEIGHTS
YES
CHECK APPLICATION PRESSURE
OK
NO
C2
ADJUST UNTIL CORRECT
RUN & RE-CHECK WEIGHTS
YES
DUMP GLUE, REFILL & RE-CHECK
OK
NO
C3
CHECK DISPENSE TIME = 30'S
OK
NO
ADJUST SENSOR UNTIL CORRECT
YES
RUN
YES
CHECK HEATS TO STANDARD
OK
NO
C4
ADJUST UNTIL CORRECT
RUN & RE-CHECK WEIGHTS
YES
CHECK SOLVENTLESS TO STANDARD
OK
NO
C5
ADJUST UNTIL CORRECT
RUN & RE-CHECK WEIGHTS
YES
CALL FOR HELP
R REES 4/10/2006

The flow diagram above is a simple process audit for the 5 variables under the control of the machine operator to control glue weights in the manufacture of polythene food wrapping.

The important element of this standard fault-finding procedure is that it gives the operator permission to fail. It does not give them permission to adjust away from standard to get around a problem. Once they have checked these 5 settings, what do we know is definitely not the problem? These 5 variables! so why should we move them? Take your hands-off call for help and let's find out what is really wrong as it could be something that is very important.

The mechanical condition of the machine could be deteriorating, and you don't want to get around this problem with non-standard settings. You want to find the real cause and fix it. Of course this may not be possible immediately, you may have to wait until a natural downtime period. Non-standard settings in the meantime are OK. But once to fault is corrected, then standard settings should be returned.

Using control charts in the correct way like this has one main effect on your processes; it will make you keep your hands off the process adjustments more. Avoiding un-necessary process adjustments that would make the defect rate worse. S.P.C. used correctly will make your defect rates go down!

The less you touch a process the better it gets...

One last point about keeping your hands off a random patterned process. Everyone says 'but I have to adjust' there's a pattern. I have to re-set the process as it wears.

This type of process performance pattern is characterised below. And this is the only situation were the more often you 'interfere' the more you can reduce the process variability. In this case the machine is being re-set every 10 pieces. If it was re-set ever 5 then the variability would be halved.

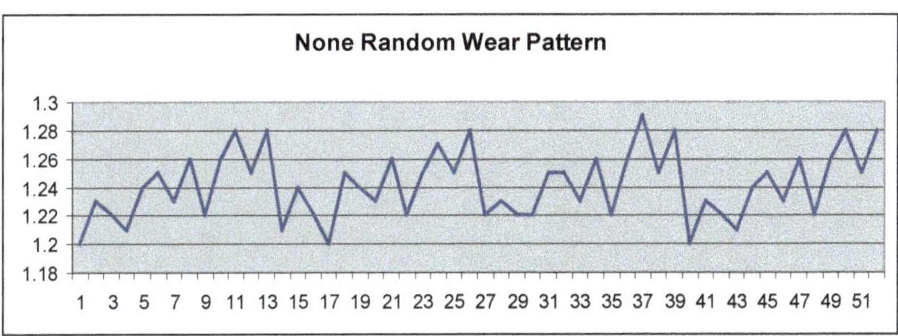

However, for this 're-setting' to be acceptable you need to see this 'saw tooth' pattern on a chart. I would guess over the 25 years I've been involved in process improvement work, 80% of the time that a client claims this pattern exists, when a simple run is plotted, the pattern is random and not moving as thought.

When your process is a genuine random number generator SPC is an absolute must. It will make your operators and technicians keep their hands in their pockets more (drink tea and read the paper!) avoiding making the defect rates worse. From observation they say that at least 85% of the time adjustment makes the process worse! Once again here is a statistical tool to help you save time and money and make better process management decisions.

Summary.

When you reach this point in your project or process journey..

CHAOS to CONTROL.

You've pretty much arrived at a controlled process where all the variables are linked to one of the control systems of work used on your site. You've worked your way through this diagram and the D.M.A.I.C steps, using appropriate tools to understand, measure and control your process physics.

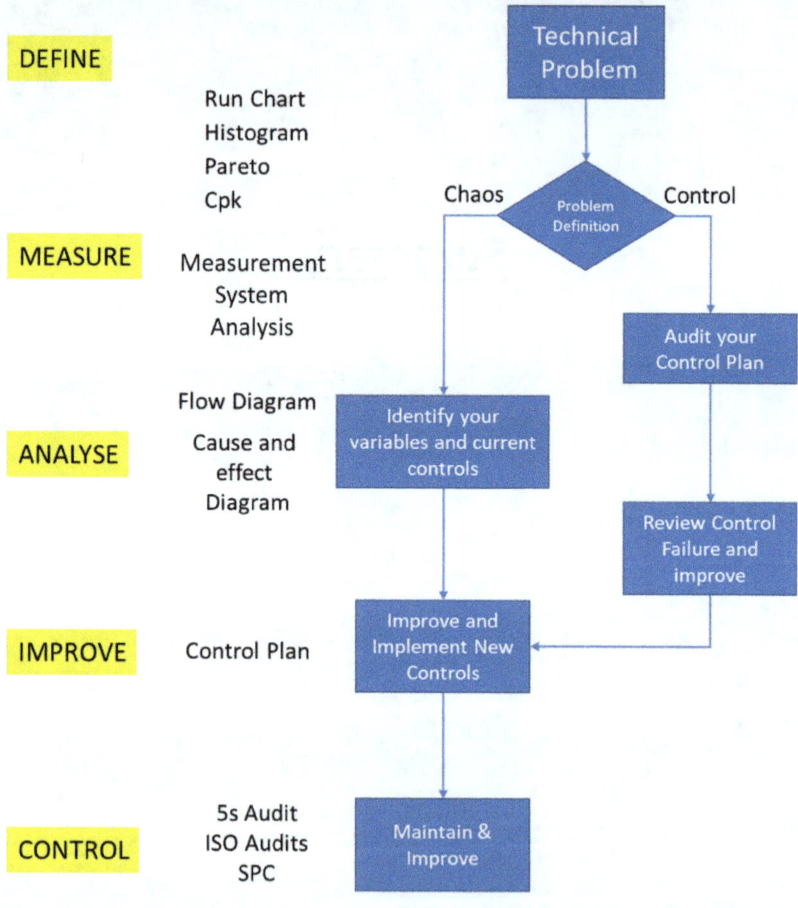

Remember to always think statistically and use Process Physics as your guide. Look at your process problem and decide if you need to control Noise (Chaos) or understand the Signal

(Control). Statistically, you're asking do I need to change the spread of the results (Standard Deviation) or the centre of the results (Mean). Get your hands on variable data and look at your process through the lens of simple diagrams. Run charts, histograms, Cpk diagrams, Pareto's etc. Once you know what the problem is, Standard Deviation or Mean. Choose the tools to solve the problem, process flow, cause and effect. Identify every input variable, list every control you have in place and then decide what to do next. More controls to get rid of the Chaos? or more analysis to understand best settings?

But when you've improved your process to a satisfactory level you should see that..

Your SOP's control inputs (Fixed never to move)
Your 5s controls inputs (Fixed never to move)
TPM controls inputs (Fixed never to move)
ISO 9000 audits, audit process inputs.
SPC controls and audits inputs

If you do this, every system you have is focussed on pleasing customers and making money. Ask yourself currently is 5s focussed on making money? ISO 9000, is it focussed on making money or putting a certificate on the wall? Are your controls systems disconnected from the money making process as shown below?? Or....

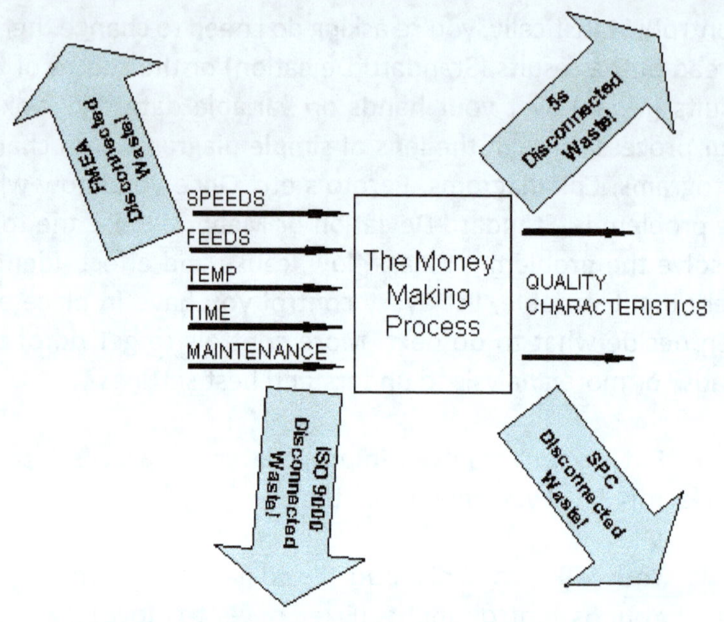

Look at the Diagram below if you can get your manufacturing business set up like this. Not only will you have World class teams who can solve any problem. You'll have a world class defect free business.

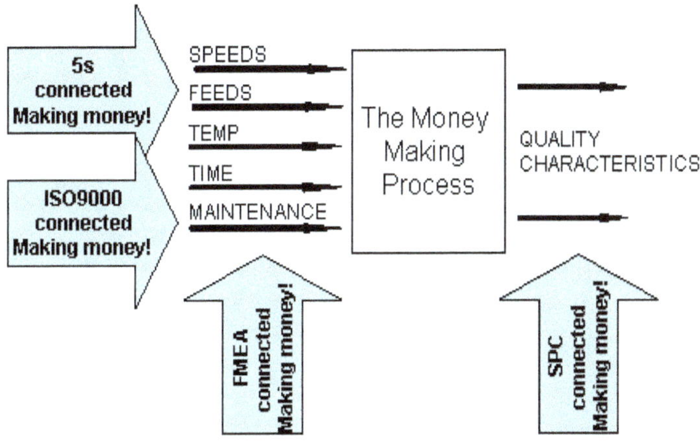

If you enjoyed this book and you think someone else would be interested in the principles in this book. Please direct them to my website www.allenp.co.uk or to Lulu.com where they can buy a copy.

If you want help implementing these principles and tools in your company or you just need some simple advice ...contact me on

Paul.allen@allenp.co.uk – 07960 053947

And finally remember get your very expensive highly technical process variables under control, find the best settings. Dial them in then sit back....DRINK TEA AND READ THE PAPER!!

www.ingramcontent.com/pod-product-compliance
Lightning Source LLC
Chambersburg PA
CBHW051502170526
45166CB00001B/349

* 9 7 8 1 4 7 1 0 2 0 1 4 8 *